797,885 Books

are available to read at

Forgotten Books

www.ForgottenBooks.com

Forgotten Books' App
Available for mobile, tablet & eReader

ISBN 978-1-332-30066-2
PIBN 10311191

This book is a reproduction of an important historical work. Forgotten Books uses
state-of-the-art technology to digitally reconstruct the work, preserving the original format
whilst repairing imperfections present in the aged copy. In rare cases, an imperfection in
the original, such as a blemish or missing page, may be replicated in our edition. We do,
however, repair the vast majority of imperfections successfully; any imperfections that
remain are intentionally left to preserve the state of such historical works.

Forgotten Books is a registered trademark of FB &c Ltd.
Copyright © 2015 FB &c Ltd.
FB &c Ltd, Dalton House, 60 Windsor Avenue, London, SW19 2RR.
Company number 08720141. Registered in England and Wales.

For support please visit www.forgottenbooks.com

1 MONTH OF FREE READING

at

www.ForgottenBooks.com

By purchasing this book you are eligible for one month membership to ForgottenBooks.com, giving you unlimited access to our entire collection of over 700,000 titles via our web site and mobile apps.

To claim your free month visit: www.forgottenbooks.com/free311191

* Offer is valid for 45 days from date of purchase. Terms and conditions apply.

Similar Books Are Available from
www.forgottenbooks.com

A First Course in Statistics
by D. Caradog Jones

Statistical Theory
by Robert Lemuel Wiggins

On the Foundations of Statistical Inference
Binary Experiments, by Allan Birnbaum

Introduction to Mathematical Statistics
by Carl Joseph West

The Mathematical Theory of Probabilities, Vol. 1
And Its Application to Frequency Curves and Statistical Methods, by Arne Fisher

On a Novel Method of Regarding the Association of Two Variates Classed Solely in Alternate Categories
by Karl Pearson

A Treatise on Probability
by John Maynard Keynes

Frequency-Curves and Correlation
by William Palin Elderton

Introduction to Economic Statistics
by George R. Davies

Statistical Averages
A Methodological Study, by Franz Zizek

Statistical Methods
With Special Reference to Biological Variation, by Charles Benedict Davenport

The Probability That a Numerical, Analysis Problem Is Difficult
by James W. Demmel

Lectures on the Theory of Industrial Sampling
Preface and Summary of the Text, by J. H. Curtiss

An Elementary Manual of Statistics
by A. L. Bowley

The Measurement of Groups and Series a Course of Lectures
by A. L. Bowley Ma

The Elements of Vital Statistics
by Arthur Newsholme

Statistics and Sociology, Vol. 1
by Richmond Mayo-Smith

The Measurement of Groups and Series
A Course of Lectures, by Arthur Lyon Bowley

Readings and Problems in Statistical Methods
by Horace Secrist Phd

Notes on Paretian Distribution Theory
by G. M. Kaufman

THE UNIVERSITY OF KANSAS
MUSEUM OF NATURAL HISTORY

SPECIAL PUBLICATION
No. 19

THE COMPLEAT CLADIST

A Primer of Phylogenetic Procedures

E. O. WILEY
D. SIEGEL-CAUSEY
D. R. BROOKS
V. A. FUNK

LAWRENCE

October 1991

The University of Kansas
Museum of Natural History
Special Publications

Museum of Comparative Zoology Library
Harvard University

To receive our 1990 Catalog of Publications, contact Publications, Museum of Natural History, The University of Kansas, Lawrence, Kansas 66045-2454, USA. To order by phone call 913-864-4540. MasterCard and VISA accepted. See the inside back cover for a list of other available numbers in this series.

The University of Kansas
Museum of Natural History

Special Publication No. 19
October 1991

THE COMPLEAT CLADIST

A Primer of Phylogenetic Procedures

E. O. Wiley
Museum of Natural History
The University of Kansas
Lawrence, Kansas 66045

D. Siegel-Causey
Museum of Natural History
The University of Kansas
Lawrence, Kansas 66045

D. R. Brooks
Department of Zoology
The University of Toronto
Toronto, Ontario M5S 1A1
CANADA

V. A. Funk
Department of Botany
National Museum of Natural History
The Smithsonian Institution
Washington, D.C. 20560

Museum of Natural History
Dyche Hall
The University of Kansas
Lawrence, Kansas
1991

THE UNIVERSITY OF KANSAS
MUSEUM OF NATURAL HISTORY

Joseph T. Collins, *Editor*
Kimberlee Wollter, *Copyediting and Design*
Kate Shaw, *Design and Typesetting*

© 1991 Museum of Natural History
The University of Kansas
Lawrence, Kansas 66045-2454, USA

Special Publication No. 19
pp. x + 1–158; 118 figures; 58 tables
Published October 1991
ISBN 0–89338–035–0

Text for this publication was produced on a Macintosh II computer in Microsoft® Word, and figures were drafted in Claris™ MacDraw® II. The publication was then designed and typeset in Aldus PageMaker® and forwarded to the printer.

PRINTED BY
THE UNIVERSITY OF KANSAS PRINTING SERVICE
LAWRENCE, KANSAS

ERRATA

The Compleat Cladist: A Primer of Phylogenetic Procedures
by
E. O. Wiley, D. Siegel-Causey, D. R. Brooks, and V. A. Funk

On p. 74 the formula for "r" should read:

$$r = \frac{g - s}{g - m}$$

On p. 107, change legend for Figure 6.17 to read "A phylogeny of hypothetical species of the genus *Mus*, ..."

The University of Kansas
Museum of Natural History
Dyche Hall
Lawrence, KS 66045-2454

Nonprofit Org.
U.S. Postage
PAID
Lawrence, KS
Permit No. 65

Museum of Comparative Zoology
Library
26 Oxford Street
Cambridge, MA 02138

" . . . but, he that hopes to be a good Angler must not onely bring an inquiring, searching, observing wit, but he must bring a large measure of hope and patience, and a love and propensity to the Art it self; but having once got and practis'd it, then doubt not but Angling will prove to be so pleasant, that it will prove like Vertue, a reward to it self."

Piscator speaking to Venator and Auceps
The Compleat Angler by Izaak Walton
The Modern Library Printing of the Fourth (1668) Edition
Random House, New York

PREFACE

In writing this workbook, we have strived to follow in the tradition of Brooks et al. (1984) of providing a guide to basic phylogenetic techniques as we now understand them. The field of phylogenetics has undergone many changes, some philosophical and some empirical, in the last 10 years. We hope to reflect some of these changes in this workbook.

The workbook is arranged in a manner roughly functional and pedagogical. The sophisticated reader, for example, might question why we spend so much time with exercises (in Chapter 2) that do not really reflect the way "real" phylogenetic analysis is performed or why so much space is given to Hennig Argumentation and Wagner algorithms (Chapter 4) when they are either not a part of modern computer algorithms or, at best, are only a starting point for finding the best hypothesis of common ancestry for the particular data analyzed. Our answer, perhaps constrained by our own histories, is that this approach seems to help at least some students learn phylogenetics.

Because we can only touch on the most basic topics and provide exercises for only a few of these, we invite the student to explore the original literature; we have cited sources in the body of the text and called attention to some papers in the "Chapter Notes and References" sections at the end of each chapter. The absence of a paper in the text or in the "Notes" section is no reflection on the worth of the paper. A compilation of all of the useful papers relating to phylogenetics is quite beyond the scope of this workbook.

In addition to the exercises, we have provided immediate feedback sections termed "Quick Quizzes." We are interested in reader opinion regarding both the exercises and the quick quizzes. We will incorporate suggestions, wherever possible, in subsequent editions.

The major title, *The Compleat Cladist,* is inspired by the title *The Compleat Angler* by Izaak Walton, a marvelous book published in many editions since 1653. Of course, this book is not "complete" or even "compleat" in the archaic sense of representing a book that teaches complete mastery of a subject. Phylogenetics is much too dynamic for a small workbook to fulfill that criterion. Rather, we take our inspiration from Walton; the compleat cladist is one who approaches the subject with energy, wonder, and joy. Unfortunately, none of us are clever enough to come up with an analogy to the "Anglers Song."

We thank the following people for their valuable comments on part or all of the earlier drafts of this workbook: the students of Biology 864 and Mike Bamshad (University of Kansas), John Hayden (University of Richmond), Debbie McLennan (University of Toronto), David Swofford (Illinois Natural History Survey), Charlotte Taylor, Richard Thomas, and Rafael Joglar (University of Puerto Rico), Wayne Maddison (Harvard University), and Arnold Kluge (University of Michigan).

Special thanks are due to David Kizirian (University of Kansas) for working through the answers to the exercises and to Kate Shaw and Kim Wollter (University of Kansas) for their editorial skills. Partial support in the form of computer hardware and software by the National Science Foundation (BSR 8722562) and the University of Kansas Museum of Natural History is gratefully acknowledged. Mistakes in interpretation and exercise answers are our responsibility, and we will be grateful for any suggestions and corrections for incorporation into future editions.

E. O. Wiley, D Siegel-Causey, D. R. Brooks, and V. A. Funk
Lawrence, Kansas; Toronto, Ontario; and Washington, D.C.
Summer, 1991

CONTENTS

PREFACE ....
CHAPTER 1: INTRODUCTION, TERMS, AND CONCEPTS ....
    Terms for Groups of Organisms ....
        Quick Quiz—Groups ....
    Terms for the Relationships of Taxa ....
        Quick Quiz—Relationships ....
    Terms for Classifications ....
        Quick Quiz—Classification ....
    Process Terms ....
    Terms for the Attributes of Specimens ....
        Quick Quiz—Characters ....
    Chapter Notes and References ....
    Quick Quiz Answers ....
CHAPTER 2: BASIC PHYLOGENETIC TECHNIQUES ....
        Quick Quiz—Basic Rules of Analysis ....
    Sample Analyses ....
    Exercises ....
    Chapter Notes and References ....
    Quick Quiz Answers ....
CHAPTER 3: CHARACTER ARGUMENTATION AND CODING ....
    Outgroup Comparison ....
    Polarity Decisions ....
    Rules of Thumb ....
    Other Situations ....
        Quick Quiz—Outgroups and Polarities ....
    Polarity Exercises ....
    Character Coding ....
        Quick Quiz—Character Coding ....
    Coding Exercises ....
    Chapter Notes and References ....
    Quick Quiz Answers ....
CHAPTER 4: TREE BUILDING AND OPTIMIZATION ....
    Hennig Argumentation ....
        Hennig Exercises ....

The Wagner Algorithm	47
Wagner Definitions	49
The Algorithm	50
Wagner Tree Exercises	54
Optimal Trees and Parsimony Criteria	54
Optimizing Trees	56
ACCTRAN	57
ACCTRAN Exercises	60
Discussion	61
Finding MPR Sets	62
DELTRAN	63
DELTRAN Exercises	63
Current Technology	66
Chapter Notes and References	68
CHAPTER 5: TREE COMPARISONS	71
Summary Tree Measures	71
Tree Length	71
Consistency Indices	72
Ensemble Consistency Indices	75
The F-Ratio	76
Tree Summaries Exercises	78
Consensus Techniques	80
Strict Consensus Trees	81
Adams Trees	83
Majority Consensus Trees	88
Chapter Notes and References	89
CHAPTER 6: CLASSIFICATION	91
Evaluation of Existing Classifications	92
Logical Consistency	92
Determining the Number of Derivative Classifications	99
Classification Evaluation Exercises	100
Constructing Phylogenetic Classifications	102
Rules of Phylogenetic Classifications	102
Conventions	103
Quick Quiz—Taxonomy vs. Systematics	108
Convention Exercises	108
Chapter Notes and References	111
Quick Quiz Answers	111

CHAPTER 7: COEVOLUTIONARY STUDIES 113
   Coding Phylogenetic Trees 113
      Quick Quiz—Biogeography 115
      Single Tree Exercises 117
      More Than One Group 118
      Missing Taxa 120
      Widespread Species 124
      Sympatry within a Clade 127
   The Analogy between Phylogenetics and Historical Biogeography 127
   Chapter Notes and References 128
   Quick Quiz Answers 128
LITERATURE CITED 129
ANSWERS TO EXERCISES 137
   Chapter 2 137
   Chapter 3 139
   Chapter 4 141
   Chapter 5 150
   Chapter 6 151
   Chapter 7 156

Chapter 1

INTRODUCTION, TERMS, AND CONCEPTS

The core concept of phylogenetic systematics is the use of derived or apomorphic characters to reconstruct common ancestry relationships and the grouping of taxa based on common ancestry. This concept, first formalized by Hennig (1950, 1966), has been slowly, and not so quietly, changing the nature of systematics. Why should we be interested in this approach? What about phylogenetic systematics is different from traditional systematics? The answer is simple: classifications that are not known to be phylogenetic are possibly artificial and are, therefore, useful only for identification and not for asking questions about evolution.

There are two other means of making statements of relationship: traditional systematics and phenetics. Traditional systematic methods employ intuition. In practical terms, intuition is character weighting. The scientist studies a group of organisms, selects the character(s) believed to be important (i.e., conservative), and delimits species and groups of species based on these characters. Disagreements usually arise when different scientists think different characters are important. It is difficult to evaluate the evolutionary significance of groups classified by intuition because we do not know why they were created or whether they represent anything real in nature. Because these groups may not be defined at all or may be defined by characters that have no evolutionary significance, such groups may be artificial.

Phenetics is an attempt to devise an empirical method for determining taxonomic relationships. In practice, phenetics is no better than traditional systematics in determining relationships because the various algorithms concentrate on reflecting the total similarity of the organisms in question. Organisms that appear to be more similar are grouped together, ignoring the results of parallel or convergent evolution and again creating possibly artificial groups.

Phylogeneticists differ from traditional systematists in that we employ empirical methods to reconstruct phylogenies and strictly evolutionary principles to form classifications rather than relying on intuition or authority. We differ from pheneticists in that our methods seek to find the genealogic relationships among the taxa we study rather than the phenetic or overall similarity relationships.

What all this means is that the groups we discover are thought to be natural, or monophyletic. Given any array of taxa, which two are more closely related to each other than either is to any other taxon? We attempt to discover the common ancestry relationships indirectly through finding evidence for common ancestry. This evidence comes in the form of shared derived characters (synapomorphies). For example, among Aves (birds), Crocodylia (alligators and crocodiles), and Squamata (lizards, snakes, and amphisbaenians), Aves and Crocodylia are thought to be more closely related because they share a number of synapomorphies thought to have originated in their common ancestor, which appeared after (later than) the common ancestor of all three taxa. This relationship is shown in the form of a phylogenetic tree, a reconstruction of the genealogic relationships.

In addition, phylogeneticists view the reconstructed tree (frequently termed a cladogram) as the classification, and when expressing it in a hierarchical scheme, we insist on maintaining monophyletic groups and sister-group relationships. The discovery of monophyletic groups is the basic quest of phylogenetics. Going to all the trouble of finding the groups and then throwing them away does not make sense to us.

Ever since the general theory of evolution gained acceptance, systematists have sought the one evolutionary history for organisms and have tried to fit that history into a hierarchical structure. We seek to reflect in our classifications the groups that we find in nature. Because phylogenetic reasoning delimits groups based on common ancestry, we can attempt to reconstruct evolutionary histories and from them develop a hierarchical ranking scheme. Phylogenetic groups are then a reflection of the order in nature. Therefore, our classifications can be used for the study of other characters and for further investigations in biogeography, coevolution, molecular evolution, rates of evolution, ecology, etc. If you wish to use classifications to study evolution, they must reflect the genealogy of the taxa in question. Groups that are potentially artificial cannot be used in such investigations.

One of the greatest strengths of the phylogenetic system is that the method and results are transparent, meaning that decisions, whether right or wrong, are based on data that can be examined by any and all persons willing to understand the nature of the data. The phylogenetic system does not depend on some special and mysterious knowledge about organisms that only the "expert" can understand. A critic cannot claim that your idea of the phylogenetic history of a group is wrong just because he has studied the groups longer than you have. Of course, there are valid disagreements, and there is room for change and improvement. But these disagreements are data based, not opinion based. Phylogenetics, to put it crudely, is a put-up-or-shut-up scientific discipline.

This workbook presents the basics of phylogenetic systematics as we use it today. We also cite references for those interested in following some of the debates currently underway among the proponents of phylogenetic systematics. We hope that this information will stimulate you and illustrate the importance of systematics as the basis of comparative biology. When you have finished this workbook, you should be able to reread this introduction and understand what we are trying to accomplish. As an acid test, go read Hennig (1966); it's the way we got started, and it remains the classic in the field.

All new scientific ideas and analytical methods are accompanied by new sets of terms and concepts, which can be unsettling to the tyro and even more unsettling to the experienced systematist who is called upon to abandon the "traditional" meanings of terms and embrace new meanings. The basic rationale for adopting the definitions and concepts presented in this workbook is twofold. First, it is vitally important for systematics and taxonomy to be integrated into in the field of evolutionary theory. Willi Hennig's major motivation for reforming systematics and taxonomy was to bring them in line with the Darwinian Revolution, making the results obtained through phylogenetic systematics directly relevant to studies in other fields of evolutionary research. Second, it is vitally important that the terms used in an empirical field be as unambiguous as possible so that hypotheses are as clear as possible. With these rationales in mind, we offer the following definitions for the basic terms in our field. They are largely taken from Hennig (1966) or Wiley (1980, 1981a). Other more specialized terms will be introduced in other chapters.

Terms for Groups of Organisms

1. Taxon.—A **taxon** is a group of organisms that is given a name. The name is a proper name. The form of many of these proper names must follow the rules set forth in one of the codes that govern the use of names. The relative hierarchical position of a taxon in a classification can be indicated in many ways. In the Linnaean system, relative rank is denoted by the use of categories. *You should not confuse the rank of a taxon with its reality as a group.* Aves is a taxon that includes exactly the same organisms whether it is ranked as a class, an order, or a family.

2. Natural taxon.—A **natural taxon** is a group of organisms that exists in nature as a result of evolution. Although there are many possible groupings of organisms, only a few groupings comprise natural taxa. In the phylogenetic system, there are two basic kinds of natural taxa: species and monophyletic groups. A **species** is a lineage. It is a taxon that represents the largest unit of taxic evolution and is associated with an array of processes termed speciation. A **monophyletic group** is a group of species that includes an ancestral species and all of its descendants (Fig. 1.1a). Members of monophyletic groups share a set of common ancestry relationships not shared with any other species placed outside the group. In other terms, a monophyletic group is a unit of evolutionary history. Examples include Mammalia and Angiospermae.

3. Clade.—A **clade** is a monophyletic group, i.e., a natural taxon.

4. Ancestral taxon.—An **ancestral taxon** is a species that gave rise to at least one new daughter species during speciation, either through cladogenesis or reticulate speciation. By cladogenesis we mean speciation that results in two or more branches on the phylogenetic tree where there was only one branch before. By reticulate speciation we mean the establishment of a new species through a hybridization event involving two different species. A species that emerged from cladogenesis has one ancestral species but a species emerging from reticulate speciation has two ancestral species. In the phylogenetic system, only species can be ancestral taxa. Groups of species are specifically excluded from being ancestral to other groups of species or to single species. The biological rationale for this distinction is clear; there is an array of processes termed speciation that allow for one species to give rise to another (or two species to give rise to a species of hybrid origin), but there are no known processes that allow for a genus or a family to give rise to other taxa that contain two or more species ("genusation" and "familization" are biologically unknown). Thus, each monophyletic group begins as a single species. This species is the ancestor of all subsequent members of the monophyletic group.

5. Artificial taxon.—An **artificial taxon** is one that does not correspond to a unit involved in the evolutionary process or to a unit of evolutionary history. You will encounter

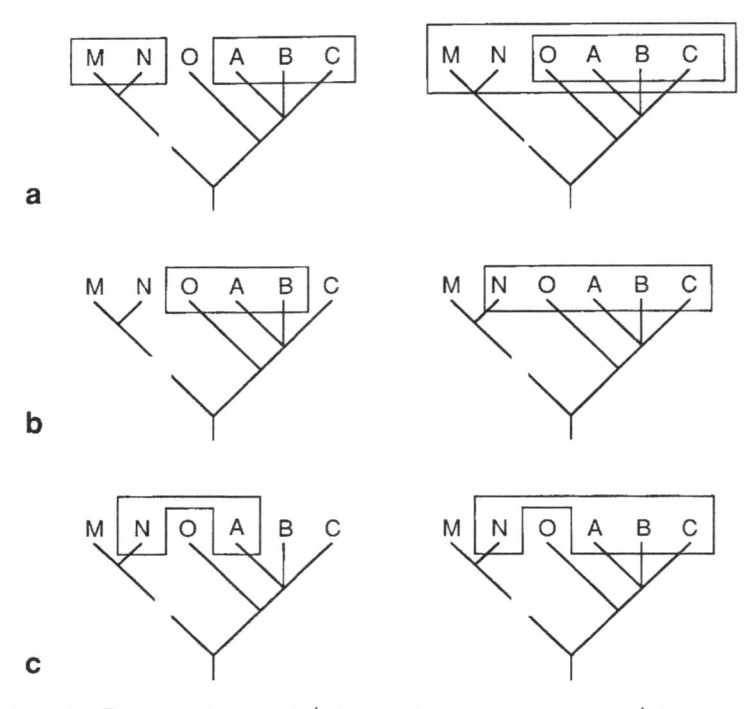

Fig. 1.1.—Examples of monophyletic (a) paraphyletic (b), and polyphyletic (c) groups.

two kinds of artificial groups. **Paraphyletic groups** are artificial because one or more descendants of an ancestor are excluded from the group (Fig. 1.1b). Examples include Dicotyledonae, Vermes, and Reptilia. **Polyphyletic groups** are artificial because the common ancestor is placed in another taxon (Fig. 1.1c). An example would be the Homeothermia, a group composed of birds and mammals. Note that the term "ancestor" is used in its logical sense, i.e., the ancestor is unknown but its inclusion or exclusion can be deduced as a logical consequence of the grouping. The important contrast is between monophyletic groups and nonmonophyletic groups. Paraphyletic groups are as artificial as polyphyletic groups. Further, it is not always possible to distinguish clearly the status of a group as either paraphyletic or polyphyletic.

6. Grade.—A **grade** is an artificial taxon. Grade taxa are frequently paraphyletic and sometimes polyphyletic but are supposed to represent some level of evolutionary progress, level of organization, or level of adaptation (e.g., Reptilia or Vermes).

7. Ingroup.—The **ingroup** is the group actually studied by the investigator (Fig. 1.2a). That is, it is the group of interest.

8. Sister group.—A **sister group** is the taxon that is genealogically most closely related to the ingroup (Fig. 1.2a). The ancestor of the ingroup cannot be its sister because the ancestor is a member of the group.

INTRODUCTION, TERMS, AND CONCEPTS

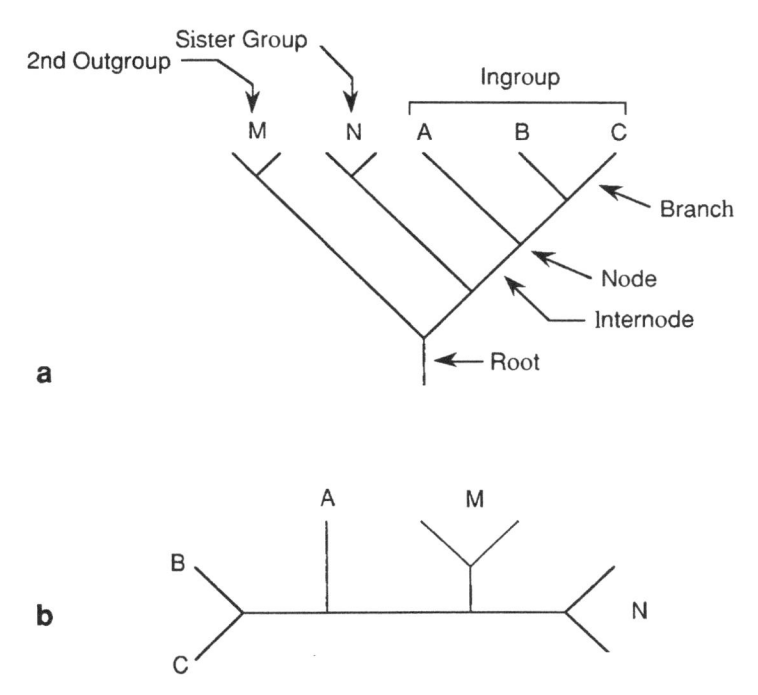

Fig. 1.2.—A rooted (a) and unrooted (b) tree for the group ABC and two of its outgroups, N (the sister group) and M.

9. Outgroup.—An **outgroup** is any group used in an analysis that is not included in the taxon under study. It is used for comparative purposes, usually in arguments concerning the relative polarity of a pair (or series) of homologous characters. The most important outgroup is the sister group, and considerable phylogenetic research may be needed to find the sister group. Usually more than one outgroup is needed in an analysis. This will become apparent in Chapter 3.

Quick Quiz—Groups

Examine Fig. 1.1 and answer the following:
1. Why do we say that the group A+B+C and the group M+N are monophyletic?
2. Which taxa would have to be either included or excluded to change the paraphyletic groups into monophyletic groups?
3. Can polyphyletic groups ever contain monophyletic groups within them?
4. Where are the ancestors in these diagrams?

Terms for the Relationships of Taxa

1. Relationship.—In the phylogenetic system, the term **relationship** refers to the genealogic or "blood" relationship that exists between parent and child or between sister and brother. In other systems, relationship can also refer to similarity, with the evolutionary implication that taxa that are more similar to each other are more closely related. *This meaning is specifically excluded from the phylogenetic system.*

2. Genealogy and genealogic descent.—A **genealogy** is a graphic representation of the descent of offspring from parents. **Genealogic descent** on the taxon level (i.e., between groups recognized as taxa) is based on the proposition that species give rise to daughter species through an array of mechanisms termed speciation.

3. Tree.—A **tree** is a branching structure and, in our sense, might contain reticulations as well as branches. A tree may be rooted (Fig. 1.2a) or unrooted (Fig. 1.2b) and is composed of several parts. A **branch** is a line connecting a branch point to a terminal taxon. A branch point, or **node**, represents a speciation event. This is true even if the taxa joined by the branch point are higher taxa such as families or phyla, because higher taxa originated as species. Branch points are sometimes represented by circles. An **internode** is a line connecting two speciation events and represents at least one ancestral species. (We say at least one because the statement is made relative to the species and groups we actually know about. It is always possible to find a new species or group of species that belongs to this part of the phylogeny. To make this addition, we would bisect the internode and create the possibility for an additional ancestral species.) The internode at the bottom of the tree is given the special term **root**. The term **interval** is a synonym of internode and is used in the Wagner algorithm (see Chapter 4). A **neighborhood** is an area of a tree relative to a particular taxon or taxa. In Fig. 1.2b, taxon B is the **nearest neighbor** of taxa A and C. Note that A may or may not be the sister of a monophyletic group B+C. This relationship cannot be established until the root is specified.

4. Phylogenetic tree.—A **phylogenetic tree** is a graphic representation of the genealogic relationships between taxa as these relationships are understood by a particular investigator. In other words, a phylogenetic tree is a hypothesis of genealogic relationships on the taxon level. *Although it is possible for an investigator to actually name ancestors and associate them with specific internodes, most phylogenetic trees are common ancestry trees.* Further, phylogenetic trees are hypotheses, not facts. Our ideas about the relationships among organisms change with increasing understanding.

5. Cladogram.—**Cladograms** are phylogenetic trees. They have specific connotations of implied ancestry and a relative time axis. Thus, a cladogram is one kind of phylogenetic tree, a common ancestry tree. In some modifications of the phylogenetic system, specifically what some have termed Transformed Cladistics, the cladogram is the basic unit of analysis and is held to be fundamentally different from a phylogenetic tree. Specifically, it is purely a depiction of the derived characters shared by taxa with no necessary connotation of common ancestry or relative time axis.

6. Venn diagram.—A **Venn diagram** is a graphic representation of the relationships among taxa using internested circles and ellipses. The ellipses take the place of internode connections. A typical Venn diagram is contrasted with a phylogenetic tree in Fig. 1.3.

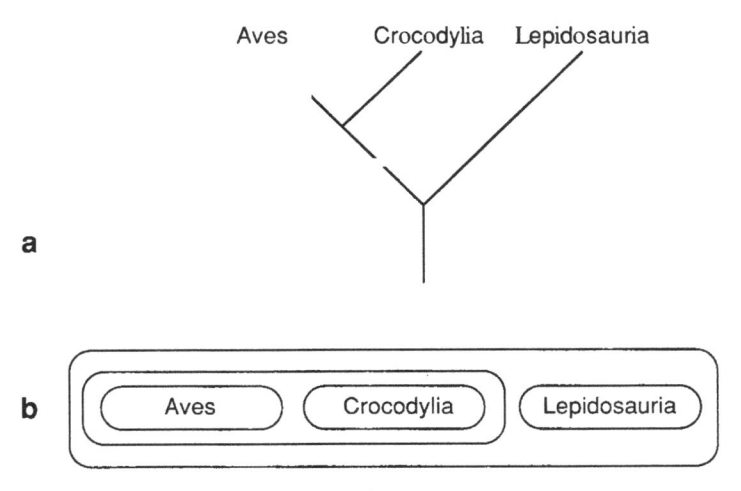

Fig. 1.3.—A phylogenetic tree (a) and a Venn diagram (b) of three groups of tetrapod vertebrates.

Quick Quiz—Relationships

Examine Fig. 1.2a and answer the following:
1. What is the sister group of the clade N?
2. What is the sister group of the clade M?
3. What is the sister group of a group composed of M+N?
4. Where is the hypothetical ancestor of the ingroup on the tree?
5. How many ancestors can a group have?
6. Draw a Venn diagram of Fig. 1.2a.

TERMS FOR CLASSIFICATIONS

1. Natural classification.—A classification containing only monophyletic groups and/or species is a **natural classification**. A natural classification is logically consistent with the phylogenetic relationships of the organisms classified as they are understood by the investigator constructing the classification. That is, the knowledge claims inherent in a natural classification do not conflict with any of the knowledge claims inherent in the phylogenetic tree. The protocols for determining if a classification is logically consistent with a phylogenetic tree are given in Chapter 6.

2. Artificial classification.—An **artificial classification** is a classification containing one or more artificial groups (i.e., one or more paraphyletic or polyphyletic groups). An artificial classification is logically inconsistent with the phylogenetic relationships of the organisms as they are understood by the investigator making the classification. That is, some of the knowledge claims inherent in the classification conflict with knowledge claims in the phylogenetic tree.

3. Arrangement.—An **arrangement** is a classification of a group whose phylogenetic relationships are not known because no investigator has ever attempted to reconstruct the evolutionary history of the group. The vast majority of current classifications are arrangements. A particular arrangement may turn out to be either a natural or an artificial classification. Arrangements serve as interim and completely necessary vehicles for classifying organisms until the phylogenetic relationships of these organisms can be worked out.

4. Category.—The **category** of a taxon indicates its relative place in the hierarchy of the classification. The Linnaean hierarchy is the most common taxonomic hierarchy and its categories include class, order, family, genus, and species. The formation of the names of taxa that occupy certain places in the hierarchy are governed by rules contained in various codes of nomenclature. For example, animal taxa ranked at the level of the category family have names that end in -idae, whereas plant taxa ranked at this level have names that end in -aceae. It is important to remember that the rank of a taxon does not affect its status in the phylogenetic system. To the phylogeneticist, all monophyletic taxa are equally important and all paraphyletic and polyphyletic taxa are equally misleading.

Classifications and arrangements are usually presented as hierarchies of names, with relative position in the hierarchy (rank) noted by categories. However, these classifications can be portrayed as tree diagrams and as Venn diagrams. The use of these methods of presenting classifications is discussed in Chapter 6.

Quick Quiz—Classification

1. In the phylogenetic system, must the taxa be clades?
2. In the phylogenetic system, must categories be clades?
3. Which is more important, a phylum or a genus?

PROCESS TERMS

Three process terms are of particular importance in the phylogenetic system. **Speciation** results in an increase in the number of species in a group. Speciation is not a single process but an array of processes. **Cladogenesis** is branching or divergent evolution and is caused by speciation. **Anagenesis** is change within a species that does not involve branching. The extent to which anagenesis and cladogenesis are coupled is an interesting evolutionary question but not a question that must be settled to understand the phylogenetic system.

INTRODUCTION, TERMS, AND CONCEPTS

TERMS FOR THE ATTRIBUTES OF SPECIMENS

1. Character.—A **character** is a feature, that is, an observable part of, or attribute of, an organism.

2. Evolutionary novelty.—An **evolutionary novelty** is an inherited change from a previously existing character. The novelty is the homologue of the previously existing character in an ancestor/descendant relationship. As we shall see below, novelties are apomorphies at the time they originate.

3. Homologue.—Two characters in two taxa are **homologues** if one of the following two conditions are met: 1) they are the same as the character that is found in the ancestor of the two taxa or 2) they are different characters that have an ancestor/descendant relationship described as preexisting/novel. The ancestral character is termed the **plesiomorphic character**, and the descendant character is termed the **apomorphic character**. The process of determining which of two homologues is plesiomorphic or apomorphic lies at the heart of the phylogenetic method and is termed **character polarization** or **character argumentation**. Three (or more) characters are homologues if they meet condition 2.

4. Homoplasy.—A **homoplasy** is a similar character that is shared by two taxa but does not meet the criteria of homology. Every statement of homology is a hypothesis subject to testing. What you thought were homologues at the beginning of an analysis may end up to be homoplasies.

5. Transformation series.—A **transformation series** (abbreviated TS in some tables and exercises) is a group of homologous characters. If the transformation series is ordered, a particular path of possible evolution is specified but not necessarily the direction that path might take. All transformation series containing only two homologous characters (the binary condition) are automatically ordered but not necessarily polarized (contrast Fig. 1.4a and Fig. 1.4b). Transformation series having more than two characters are termed multicharacter (or multistate) transformation series. If a multistate transformation series is unordered (Fig. 1.4c), several paths might be possible. Ordered transformation series are not the same as polarized transformation series (compare Figs. 1.4d and 1.4e). An **unpolarized transformation series** is one in which the direction of character evolution has not been specified (Figs. 1.4a, c, d). A **polarized transformation series** is one in which the relative apomorphy and plesiomorphy of characters has been determined by an appropriate criterion (Figs. 1.4b, e). It is possible for a transformation series to be both unordered and polarized. For example, we might know from outgroup comparison that 0 is the plesiomorphic state, but we might not know whether 1 gave rise to 2, or vice versa, or whether 1 and 2 arose independently from 0. Ordering and polarization of multicharacter transformation series can become very complicated, as we shall see in Chapter 3. Our use of the convention "transformation series/character" differs from that of many authors who use "character" as a synonym for "transformation series" and "character state" as a synonym for "character." We use "transformation series/character" instead of "character/character state" in our research and in this workbook for philosophical reasons. The "character/character state" convention reduces "character" to a term that does not refer to the attributes of organisms but instead to a class construct that contains the attributes of organisms, homologues or not. For example,

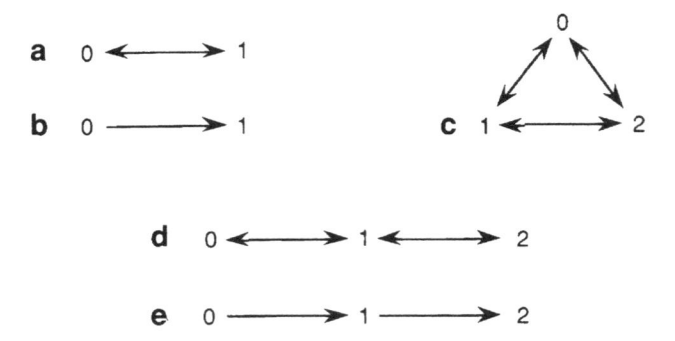

Fig. 1.4.—Characters. a. Unpolarized binary characters. b. Polarized binary characters. c. An unordered transformation series of three characters. d. The same transformation series ordered but not polarized. e. The same transformation series ordered and polarized.

dandelions do not have "color of flower" as an attribute; they have "yellow flowers." We adopt "transformation series/character" because it explicitly avoids the construction of character classes and implicitly encourages the investigator to use characters hypothesized to be homologues of each other.

6. Character argumentation.—**Character argumentation** is the logical process of determining which characters in a transformation series are plesiomorphic and which are apomorphic. Character argumentation is based on *a priori* arguments of an "if, then" deductive nature and is based on outgroup comparison. This process is frequently termed "polarizing the characters." **Polarity** refers to which of the characters is plesiomorphic or apomorphic. Character argumentation will be covered in detail in Chapter 3.

7. Character optimization.—**Character optimization** consists of *a posteriori* arguments as to how particular characters should be polarized given a particular tree topology. Character optimization might seem *a priori* when used in a computer program, but it is not.

8. Character code and data matrix.—Phylogenetic systematists are quickly converting to computer-assisted analysis of their data. When using a computer, the investigator produces a **data matrix**. Usually, the columns of the matrix are transformation series and the rows are taxa. A **code** is the numerical name of a particular character. By convention, the code "1" is usually assigned to the apomorphic character and "0" to the plesiomorphic character of a transformation series if the polarity of that series is determined (=hypothesized) by an appropriate method of polarization. If the transformation series consists of more than two characters, additional numerical codes are assigned. Alternatively, the matrix might be coded using binary coding as discussed in Chapter 3. There are many ways of reflecting the code of a character when that character is placed on a tree. We will use the following convention: characters are denoted by their transformation series and their code. The designation 1-1 means "transformation series 1, character coded 1." Some basic ways of coding characters are discussed in Chapter 3.

9. Tree length.—The length of a tree is usually considered the number of evolutionary transformations needed to explain the data *given* a particular tree topology.

You will probably need some time to assimilate all of the definitions presented. A good strategy is to review this chapter and then go to Chapter 2, working your way through the examples. We have found that deeper understanding comes from actual work. Although we cannot pick a real group for you to work on, we have attempted the next best thing, a series of exercises designed to teach basic phylogenetic techniques that we hope will mimic real situations as closely as possible.

Quick Quiz—Characters

1. How would the transformation series in Fig. 1.4c look if it were polarized and unordered?
2. Is character "1" in Fig. 1.4e apomorphic or plesiomorphic?

CHAPTER NOTES AND REFERENCES

1. There is no substitute for reading Hennig (1966). We suggest, however, that you become familiar with most of the basics before attempting to read the 1966 text. Hennig (1965) is the most accessible original Hennig. Other classics include Brundin (1966) and Crowson (1970). An interesting analysis of Hennig's impact on systematics can be found in Dupuis (1984). A considerable portion of the history of phylogenetic thought (and indeed post-1950 systematics) can be followed in a single journal, *Systematic Zoology*. We highly recommend that students examine this journal.

2. Post-Hennig texts that are suitable for beginners are Eldredge and Cracraft (1980), Wiley (1981a), Ridley (1985), Schoch (1986), Ax (1987), and Sober (1988a). A more difficult text written from the point of view of the transformed cladists is Nelson and Platnick (1981).

3. A very readable review of the entire field of systematics is Ridley (1985), whose defense of phylogenetics and criticisms of traditional (evolutionary) taxonomy, phenetics, and transformed cladistics are generally on the mark.

QUICK QUIZ ANSWERS
Groups

1. They are monophyletic because no descendant of their respective common ancestor is left out of the group.
2. To make the group O+A+B monophyletic, you would have to include C. To make the group N+O+A+B+C monophyletic, you could either include M or exclude N.
3. Yes; e.g., N+O+A+B+C contains the monophyletic group ABC.
4. You were pretty clever if you answered this one because we haven't covered it yet. The ancestors are represented by internodes between branches. Obviously they are hypothetical because none of them are named.

Relationships

1. The ingroup (A+B+C) is the sister group of N.
2. N plus the ingroup is the sister group of M.
3. A group composed of M and N is paraphyletic. Paraphyletic groups are artificial and thus cannot have sister groups.
4. The internode labeled "Internode.
5. A bunch, stretching back to the origin of life. But we usually refer only to the immediate common ancestor.
6.

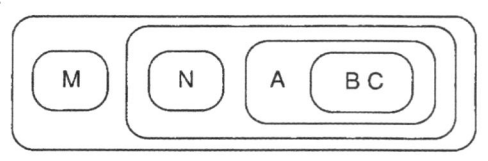

```
┌─────────────────────────────────┐
│ ┌───┐ ┌───┐ ┌───┬──────┐        │
│ │ M │ │ N │ │ A │ B C  │        │
│ └───┘ └───┘ └───┴──────┘        │
└─────────────────────────────────┘
```

Classification

1. Only clades (monophyletic groups and species) are permitted in the phylogenetic system. Grades are specifically rejected. As you will see in Chapter 6, this is because classifications that contain even a single grade are logically inconsistent with the phylogeny of the group containing the grade.
2. Categories are not taxa. They are designations of relative rank in a classification. As such, categories are neither clades nor grades.
3. All monophyletic taxa are equally important and interesting to the phylogeneticist.

Characters

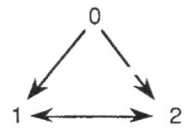

```
        0
       ╱ ╲
      ╱   ╲
     ↙     ↘
    1 ←───→ 2
```

2. Character 1 is both apomorphic and plesiomorphic. It is apomorphic relative to 0 and plesiomorphic relative to 2.

Chapter 2

BASIC PHYLOGENETIC TECHNIQUES

Phylogenetic systematists work under the principle that there is a single and historically unique genealogic history relating all organisms. Further, because characters are features of organisms, they should have a place on the tree representing this history. The proper place for a character on the tree is where it arose during evolutionary history. A "proper" tree should be one on which the taxa are placed in correct genealogic order and the characters are placed where they arose. For example, in Fig. 2.1 we show a tree of some major land plant groups with some of their associated characters. This tree can be used to explain the association of characters and taxa. The characters xylem and phloem are placed where they are because the investigator has hypothesized that both arose in the common ancestor of mosses and tracheophytes. In other words. they arose between the time of origin of the hornworts and its sister group. Xylem and phloem are thought to be homologous in all plants

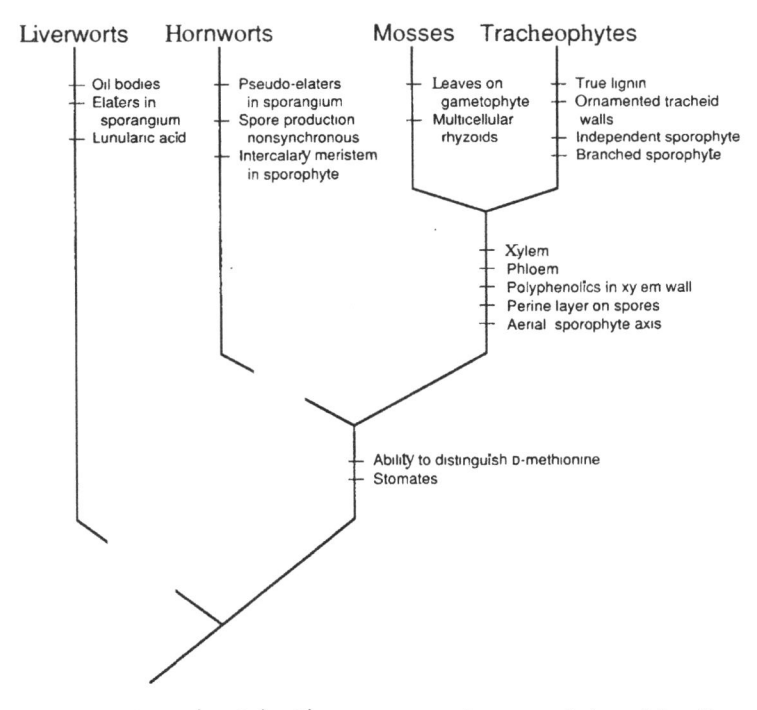

Fig. 2.1.—The phylogenetic relationships among several groups of plants (after Bremer, 1985). Synapomorphies and autapomorphies for each group are listed. Some characters are not shown.

that have these tissues. Thus, each appears only once, at the level of the tree where each is thought to have arisen as an evolutionary novelty. Now, if we did not have this tree but only the characters, we might suspect that all of the taxa that have xylem and phloem shared a common ancestor that was not shared by taxa that lack xylem and phloem. Our suspicion is based on a complex of hypotheses, including the fact that xylem and phloem from different plants are very similar, being composed of a few basic cell types or obvious derivatives of these cell types. In the phylogenetic system, such detailed similarity is always considered evidence for homology, which necessarily implies common origin. This concept is so important that it has been given the name Hennig's Auxiliary Principle (Hennig, 1953, 1966).

Hennig's Auxiliary Principle.—Never assume convergence or parallel evolution, always assume homology in the absence of contrary evidence.

This principle is a powerful one. Without it, we must pack up and go home because we could exclude any similarity that we don't wish to deal with by asserting that "it probably arose by convergent evolution." Of course, just because we use Hennig's Auxiliary Principle doesn't mean that we believe that convergences are rare or nonexistent. Convergences are facts of nature and are rather common in some groups. But to pinpoint convergence, you first have to have a tree, and without Hennig's Auxiliary Principle you will never get to a tree because you will be too worried that the characters you study are convergences. Hennig is simply suggesting that you sit back and let his method worry about convergences rather than doing something rash and ad hoc.

Back to the xylem and phloem. With Hennig's Auxiliary Principle, you can deduce that plants that have xylem and phloem shared a common ancestor not shared with other plants. Of course, you don't make such a deduction in a vacuum. You "know" that more primitive plants lack xylem and phloem, and thus it's a good guess that having xylem and phloem is more derived than lacking xylem and phloem. This deduction is a primitive sort of outgroup comparison, which we will discuss in some detail in Chapter 3. For now, we want you to consider another principle.

Grouping Rule.—Synapomorphies are evidence for common ancestry relationships, whereas symplesiomorphies, convergences, and parallelisms are useless in providing evidence of common ancestry (Hennig, 1966).

Intuitively, you know convergences and parallelisms (both termed homoplasies) are useless for showing common ancestry relationships because they evolved independently. However, plesiomorphies are also homologies. So, why can't all homologies show common ancestry relationships? The answer is that they do. It's just that symplesiomorphies can't show common ancestry relationships at the level in the hierarchy you are working at because they evolved earlier than any of the taxa you are trying to sort out. In addition, they have already been used at the level where they first appeared. If they hadn't been used, you would not be where you are. For example, you would never hypothesize that pineapples are more closely related to mosses than to some species of mistletoe based on the plesiomorphy "presence of chlorophyll." If you accepted that as valid evidence, then you would have to

BASIC PHYLOGENETIC TECHNIQUES 15

conclude that pineapples are also more closely related to green algae than to mistletoes. A common complaint by traditional taxonomists is that "cladists" use only part of the data available to them (Cronquist, 1987). This is not true, as the above example demonstrates. What we do is to attempt to find the correlation between the relative age and origin of characters.

Finally, we have to consider how to combine the information from different transformation series into hypotheses of genealogic relationships. There are several ways of accomplishing this, depending on the algorithm you use. We will find out more about this as we proceed through the workbook. For now, we will use an old-fashioned (and perfectly valid) grouping rule that goes back to the roots of the phylogenetic method, the inclusion/exclusion rule. This rule is implicit in the early work of Hennig (1966), as well as being used as an explicit rule in the much later group compatibility algorithm developed by M. Zandee (Zandee and Geesink, 1987).

Inclusion/Exclusion Rule.—The information from two transformation series can be combined into a single hypothesis of relationship if that information allows for the complete inclusion or the complete exclusion of groups that were formed by the separate transformation series. Overlap of groupings leads to the generation of two or more hypotheses of relationship because the information cannot be directly combined into a single hypothesis.

The inclusion/exclusion rule is directly related to the concept of logical consistency. Trees that conform to the rule are logically consistent with each other. Those trees that show overlap are logically inconsistent with each other. This can be shown graphically using Venn diagrams.

You can get an idea of how this rule works by studying the examples in Fig. 2.2. In Fig. 2.2a, we have four characters and four trees. The first tree contains no character information. It is logically consistent with any tree that has character information. The second tree states that N, O, and P form a monophyletic group based on characters from two transformation series (1-1 and 2-1). The third tree states that O and P form a monophyletic group based on two additional characters (3-1 and 4-1). Note that O+P is one of the possible groupings that could be found in the group N+O+P, and N+O+P completely includes O+P. The fourth tree combines these logically consistent hypotheses of relationship. Thus, these data lead to two groupings that are logically consistent with each other. The second example, Fig. 2.2b, shows the result of the inclusion of two smaller monophyletic groups (S+T) and (U+V) within a larger group (S–V). In Fig. 2.2c, we have an example of the violation of the inclusion/exclusion rule. All six transformation series imply groupings that can be included within the larger group A–D. Both C+B and C+D can be included within the group B+C+D. However, their knowledge claims conflict, and the groups overlap (Fig. 2.2d). Transformation series 1-1 and 2-1 imply a group C+B while excluding D, and transformation series 3-1 and 4-1 imply a group C+D while excluding B. C is included in two different groups, as shown by the Venn diagram in Fig. 2.2d. As a result, there are two equally parsimonious trees that are logically inconsistent with each other. To resolve which of these trees (or another tree) is preferable, we would have to analyze more data.

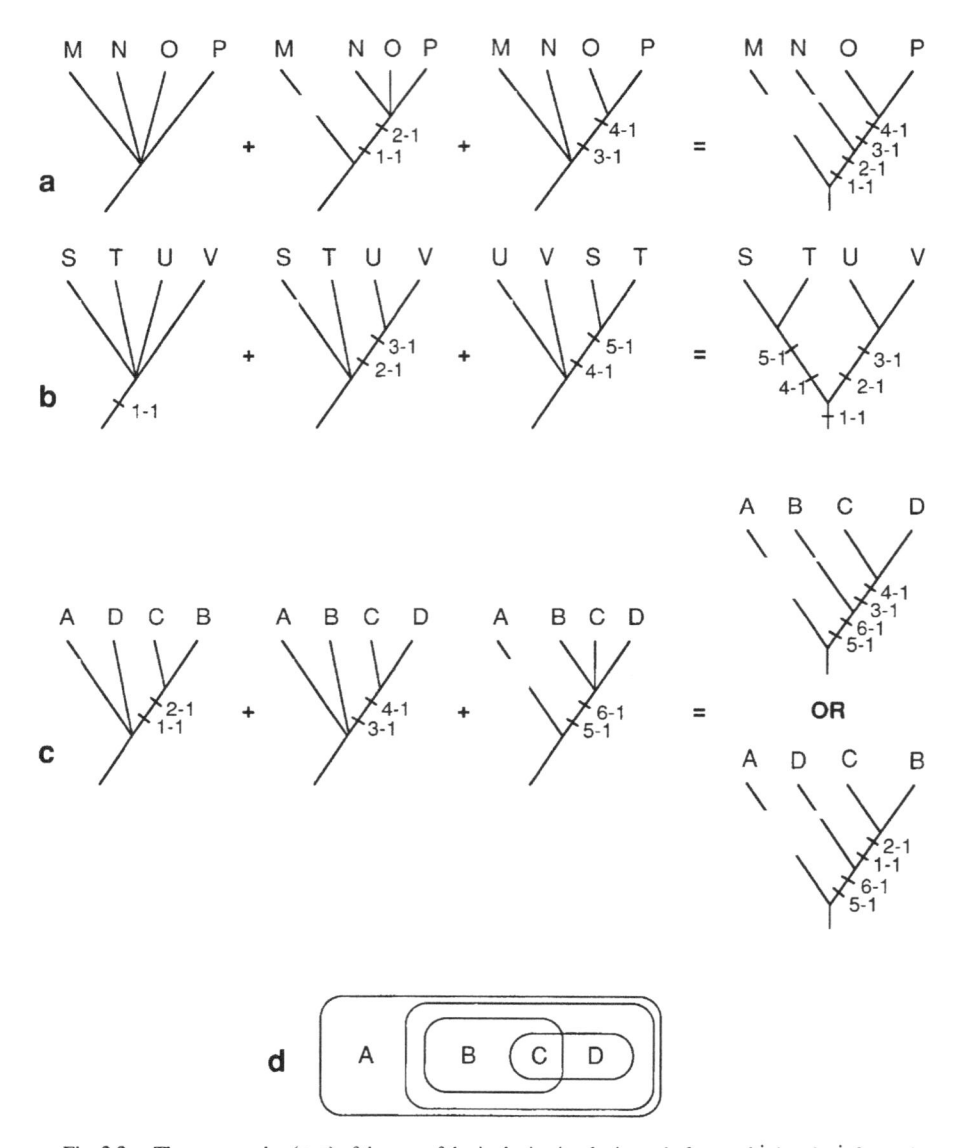

Fig. 2.2.—Three examples (a–c) of the use of the inclusion/exclusion rule for combining the information of different transformation series into trees. d. A Venn diagram showing the logical inconsistency in c.

BASIC PHYLOGENETIC TECHNIQUES 17

Quick Quiz—Basic Rules of Analysis

1. Does it follow from Hennig's Auxiliary Principle that birds and insects share a common ancestor not shared with, say, crocodiles because both have wings and are capable of flight?
2. Lizards and crocodiles have amniotic eggs. Does it follow from the Grouping Rule that lizards and crocodiles share a common ancestor?
3. How can you tell that "presence of chlorophyll a" is a plesiomorphy rather than a synapomorphy?

SAMPLE ANALYSES

We will cover the complexities of character argumentation in the next few chapters. The exercises below are based on the proposition that the outgroup has plesiomorphic characters. You can determine which character of the transformation series is plesiomorphic by simple inspection of the outgroup. (By the time you get through with Chapter 3, you will see that such a simple rule doesn't always hold, but it's good enough to get through these exercises.) Grouping is accomplished by application of the Grouping Rule. We will first take you through two exercises. Then we present a series of data matrices for you to work with. (Solutions to all exercises are in the back.)

Example 2.1.—The relationships of ABCidae.

1. Examine transformation series (TS) 1 in Table 2.1. It is composed of characters in the first column of the data matrix. We can draw a tree with the groupings implied by the synapomorphy found in the transformation series. We can do the same for TS 2. Our results look like the two trees to the left in Fig. 2.3. Because both imply the same groupings, we can say that they are topologically identical. That is, they are **isomorphic**. The combination of the two trees, by applying the Grouping Rule, is the tree on the right. We can calculate a tree length for this tree by simply adding the number of synapomorphies that occur on it. In this case, the tree length is two steps.

Table 2.1.—Data matrix for ABCidae (Example 2.1).

Taxon	Transformation series						
			3	4	5	6	
X (outgroup)	0	0	0	0	0	0	0
A	1	1	0	0	0	0	0
B	1	1	1	1	0	0	0

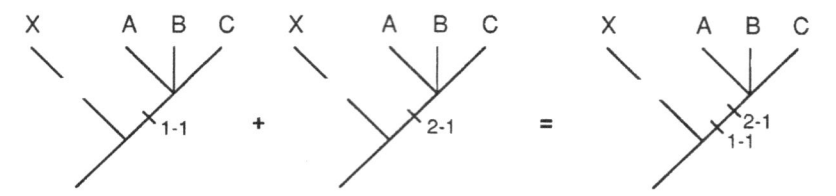

Fig. 2.3.—Trees for transformation series 1 and 2 (Example 2.1).

2. If we inspect TS 3 and TS 4, we see that the synapomorphies have identical distributions, both implying that B and C form a monophyletic group (Fig. 2.4). If we put both of these synapomorphies on a tree, the results should look like the tree on the right in Fig. 2.4. What is the length of this tree?

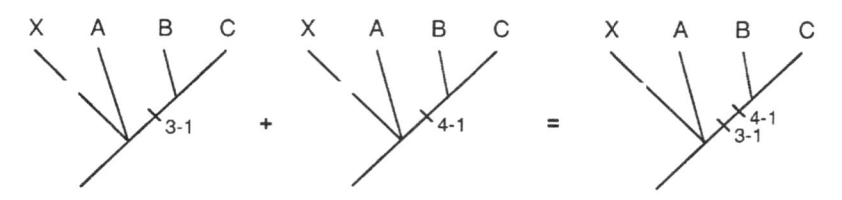

Fig. 2.4.—Trees for transformation series 3 and 4 (Example 2.1).

3. Note that only C has the apomorphies in TS 5, 6, and 7. Unique (single occurrence) apomorphies are termed **autapomorphies**. They are not useful for grouping, as we can see in Fig. 2.5, but they are useful for diagnosing C. Autapomorphies also count when figuring tree length but not when comparing trees. The length of this tree is three steps.

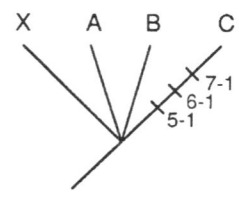

Fig. 2.5.—Tree for transformation series 5, 6, and 7 (Example 2.1).

4. We now have three different tree topologies. If we look at them closely, we can see that although the three trees are topologically different, they do not contain any conflicting information. For example, the tree implied by TS 5–7 does not conflict with the trees implied by the other transformation series because all that TS 5–7 imply is that C is different from the

BASIC PHYLOGENETIC TECHNIQUES 19

other four taxa. Further, TS 1 and 2 do not conflict with TS 3 and 4 because TS 1 and 2 imply that A, B, and C form a monophyletic group, whereas TS 3 and 4 imply that B and C form a monophyletic group but say nothing about the relationships of A or the outgroup, X. Trees that contain different but mutually agreeable groupings are **logically compatible** or **fully congruent**. They can be combined without changing any hypothesis of homology, and the length of the resulting tree is the sum of the lengths of each subtree. For example, we have combined all of the information in the data matrix to produce the tree in Fig. 2.6. Its length is seven steps, the total of the number of steps of the subtrees.

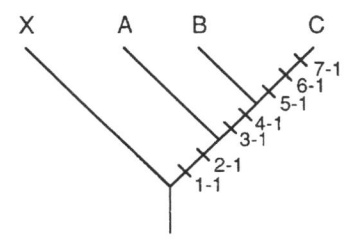

Fig. 2.6.—The best estimate of the common ancestry relationships of A B and C, given the data in Example 2.1.

Example 2.2.—Analysis of MNOidae.

The first thing you should notice about this matrix (Table 2.2) is that it has more characters scored as "1." Let's work through it.

Table 2.2.—Data matrix for MNOidae (Example 2.2).

Taxon	Transformation series						
	2	3	4	5	6		
X (outgroup)	0	0	0	0	0	0	0
M	1	1	0	0	1	1	1
N	1	1	1	1	1	1	1
O	1	1	1	1	0	0	0

1. TS 1 and TS 2 imply that M, N, and O form a monophyletic group as shown in Fig. 2.7.

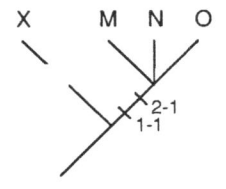

Fig. 2.7.—Tree for transformation series 1 and 2 (Example 2.2).

2. TS 3 and TS 4 imply that N and O form a monophyletic group (Fig. 2.8).

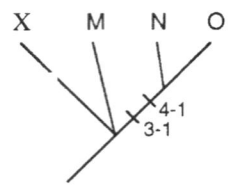

Fig. 2.8.—Tree for transformation series 3 and 4 (Example 2.2).

3. Finally, TS 5, 6, and 7 imply that M and N form a monophyletic group (Fig. 2.9).

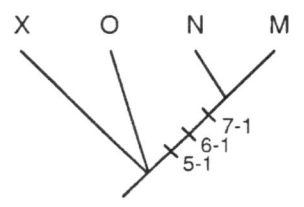

Fig. 2.9.—Tree for transformation series 5, 6, and 7 (Example 2.2).

4. At this point you should suspect that something has gone wrong. TS 3 and 4 imply a monophyletic group that includes N and O but excludes M, whereas TS 5–7 imply a monophyletic group that includes M and N but excludes O. There must be a mistake, because we have violated the inclusion/exclusion rule. In such a situation, we invoke another important principle of phylogenetic analysis: **there is only one true phylogeny.** Thus, one of our groupings must be wrong. (In fact, they might both be wrong, but the Auxiliary Principle keeps us going until such time that we *demonstrate* that both are wrong.) In this situation, we are faced with two logically incompatible trees (Fig. 2.10). Note that there is some congruence because both trees have the apomorphies of the first two transformation series.

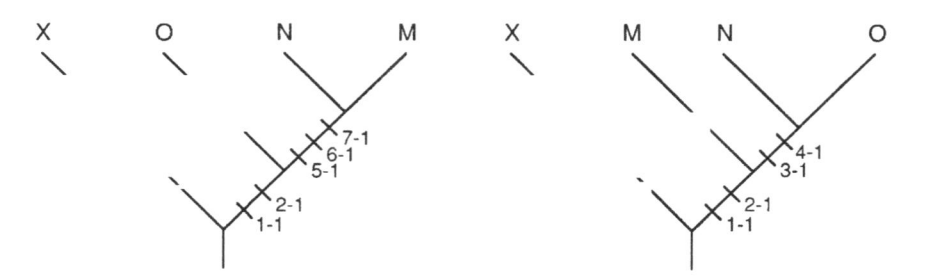

Fig. 2.10.—Trees for the two different sets of consistent transformation series (Example 2 2).

BASIC PHYLOGENETIC TECHNIQUES

5. You should have guessed by now that neither of the trees shown above is really a complete tree. The tree on the right lacks TS 5, 6, and 7, and the one on the left lacks TS 3 and 4. Leaving out characters is not acceptable. About the only way that you can get into more trouble in phylogenetic analysis is to group by symplesiomorphies. Before we start, consider how characters might be homoplasious. A character might be a convergence/parallelism, or it might be a reversal to the "plesiomorphic" character. We must consider both kinds of homoplasies. In Fig. 2.11a, TS 3 and 4 are put on the tree under the assumption that 3-1 and 4-1 arose independently (i.e., via convergence/parallelism). In Fig. 2.11b, we have placed 5-1, 6-1, and 7-1 on the alternate tree as convergences. In Fig. 2.11c, we assume that 3-1 and 4-1 arose in the common ancestor of the group and that M has reverted to the plesiomorphic character. Thus, 3-0 and 4-0 appear on the tree as autapomorphies of M. We have done the same thing for O in Fig. 2.11d for TS 5–7, given the alternative hypothesis.

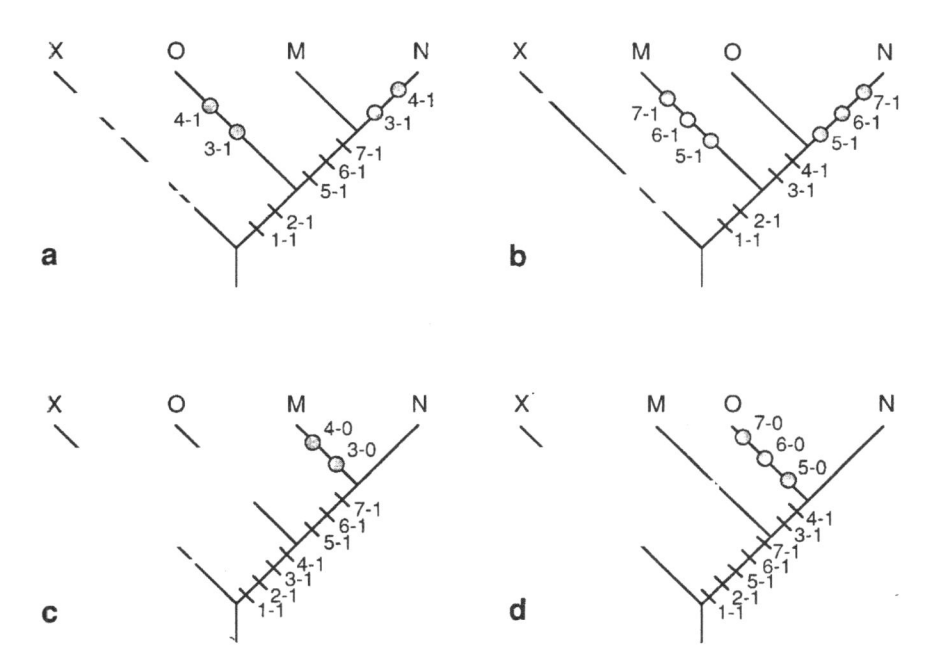

Fig. 2.11.—Alternative hypotheses of the relationships of M, N, and O based on characters of Example 2.2. ◎ = character showing convergence/parallelism or reversal (homoplasies).

6. The question is—which of these trees should we accept? That turns out to be a rather complicated question. If we adhere to the Auxiliary Principle, we should strive for two qualities, the greatest number of homologies and the least number of homoplasies. These qualities are usually consistent with each other; that is, the tree with the greatest number of synapomorphies is also the tree with the least number of homoplasies. But you can find

exceptions to this. Fortunately, both of these qualities are related to tree length. When you count the number of steps in the four trees in Fig. 2.11, you will find that trees a and c have nine steps, and trees b and d have 10 steps. We accept trees a and c as the best estimates of the phylogeny because they have a shorter length and thus the greatest number of statements of homology and the fewest number of statements of homoplasy for the data at hand. Note that such statements are *relative only to trees derived from the same data set*. The Auxiliary Principle coupled with the principle that there is only one phylogeny of life carried us to this point. Methodologically, we have employed the **principle of parsimony**. In the phylogenetic system, the principle of parsimony is nearly synonymous with the Auxiliary Principle. We can see three additional characteristics. First, trees a and c are topologically identical. Therefore, the common ancestry relationships hypothesized are identical. Second, these two trees make different claims concerning character evolution. Third, they are equally parsimonious, therefore we cannot make a choice about character evolution unless we employ some parsimony criterion other than tree length.

7. Finally, we can evaluate the performance of each character originally coded as a synapomorphy by calculating a consistency index for it. The **consistency index** (CI) of a character is simply the reciprocal of the number of times that a character appears on the tree. The CI is a favorite summary "statistic" in computer programs such as PAUP (Swofford, 1990) and MacClade (Maddison and Maddison, in press); therefore, it is good to practice some hand calculations so that you will know how the CI works. We will discuss this index and other measures of tree comparisons in Chapter 5. For example, in one most parsimonious tree (Fig. 2.11a), the apomorphy coded 1 in TS 3 appears twice, so its CI is

$$CI = \frac{1}{2} = 0.5.$$

For a given tree, we can see that a character is not really a synapomorphy by simple inspection of its CI. True homologues (real synapomorphies) have CIs of 1.0. Of course, our best estimates of true homologues come *a posteriori*, that is, in reference to the best estimate of common ancestry relationships. We do not know in advance that a particular derived similarity will turn up with a CI less than 1.0.

EXERCISES

For each of the exercises below do the following:

1. Derive trees for each transformation series or each set of transformation series with the same distribution of synapomorphies (like TS 1 and TS 2 in Example 2.1).

2. Combine the logically consistent subtrees into the shortest tree or trees accounting for all of the transformation series. Don't forget to account for the homoplasies as well as the synapomorphies and autapomorphies. In some data matrices, there will only be one such tree, in others there will be two. **Tip:** Search the trees obtained above for groups that reoccur. Use these first (for example, *sus+tus* and *vus+uus* in Exercise 2.1).

BASIC PHYLOGENETIC TECHNIQUES

3. Calculate the length of each tree and the CI for each character originally coded as a synapomorphy.

EXERCISE 2.1.—Analysis of *Sus* (Table 2.3).

Table 2.3.—Data matrix for analysis of *Sus* (Exercise 2.1).

Taxon	\multicolumn{12}{c}{Transformation series}
	1	2	3	4	5	6	7	8	9	10	11	12
Outgroup	0	0	0	0	0	0	0	0	0	0	0	0
S. sus	1	0	1	1	1	1	1	0	1	0	0	0
S. tus	1	1	1	1	1	1	0	0	0	1	0	0
S. uus	1	0	1	1	0	0	0	0	0	0	0	0
S. vus	1	1	0	0	0	0	1	1	0	0	1	0
S. wus	1	1	0	0	0	0	1	1	0	0	0	1

EXERCISE 2.2.—Analysis of Midae (Table 2.4).

Table 2.4.—Data matrix for analysis of Midae (Exercise 2.2).

Taxon	\multicolumn{11}{c}{Transformation series}
	1	2	3	4	5	6	7	8	9	10	11
Outgroup	0		0	0	0	0	0	0	0	0	0
Mus	1	1	0	1	1	1	0	0	1	1	1
Nus	1	1	0	0	1	1	1	0	0	1	1
Ous	1	1	0	0	1	1	1	0	0	1	1
Pus	0	1	1	1	0	0	0	1	0	0	0
Qus	1	1	1	1	0	1	1	0	0	0	0
Rus	1	1	1	1	0	0	1	0	0	0	0

EXERCISE 2.3.—Analysis of *Aus* (Table 2.5).

Table 2.5.—Data matrix for analysis of *Aus* (Exercise 2.3).

Taxon	\multicolumn{10}{c}{Transformation series}
	1	2	3	4	5	6	7	8	9	10
Outgroup	0	0	0	0	0	0	0	0	0	0
A. aus	1	0	1	0	1	1	0	0	0	0
A. bus	1	0	1	0	1	0	1	0	0	0
A. cus	1	1	1	0	0	0	0	0	0	1
A. dus	1	1	0	1	0	0	0	1	0	0
A. eus	1	1	0	1	0	0	1	0	1	0

CHAPTER NOTES AND REFERENCES

1. All of the texts cited in Chapter 1 cover the fundamentals of reconstructing phylogenetic relationships, but they each do so from a slightly to very different point of view. The inclusion/exclusion criterion is not usually seen in the form we present it. Indeed, it is not the way one usually goes about doing phylogenetic reconstructions. We adopted our approach because it seemed to be the simplest one to use to teach basic principles. The inclusion/exclusion approach is explicit in the "group compatibility" approach of Zandee and Geesink (1987).

2. Considerable controversy surrounds the philosophical nature of phylogenetic hypothesis testing and its relationship to parsimony. Farris (1979, 1983), Sober (1983), Kluge (1984, 1989), and Brooks and Wiley (1988) present parsimony as the relevant criterion for judging competing hypotheses. This is in direct contrast to Wiley (1975, 1981a), who attempted to reconcile parsimony and the hypothetico-deductive approach of Popper (1965), or to Felsenstein (1978, 1983), who argued that parsimony might not be the preferred criterion. Most of this controversy has been summarized by Sober (1988a).

QUICK QUIZ ANSWERS

1. Not when you examine the wings in detail. Insect wings have a completely different structure when compared with bird wings. They are so different that the best hypothesis is that the wings are not homologous. This leads to the hypothesis that flight has evolved independently in each group. Further, there is much evidence in the form of other characters that leads to the hypothesis that flying insects are more closely related to other insects that do not have wings and that birds are more closely related to other vertebrates.

2. Yes, it does follow. However, many other vertebrate groups, such as birds and mammals, also have an amniotic egg. Thus, while lizards and crocodiles certainly share a common ancestor, we cannot hypothesize that they share a common ancestor not shared also with birds and mammals. Rather, the amniotic egg is a character that provides evidence for a common ancestry relationship among all amniotes (as a synapomorphy), and its presence in lizards and crocodiles is a plesiomorphic homologous similarity.

3. You can deduce that the presence of chlorophyll a is a plesiomorphy in the same manner that you can deduce whether a character is an apomorphy, by outgroup comparison. This is covered in the next chapter.

CHAPTER 3

CHARACTER ARGUMENTATION AND CODING

This chapter is designed to teach the following skills: 1) interpretation of a phylogenetic tree in terms of nodes and internodes, 2) polarization of characters at nodes and internodes on the phylogenetic tree according to the criterion of phylogenetic parsimony as evidenced by outgroup comparison, and 3) character coding.

OUTGROUP COMPARISON

Hennig (1966) and Brundin (1966) characterized the essence of phylogenetic analysis as the "search for the sister group." They recognized that if you can find the closest relative or relatives of the group you are working on, then you have the basic tools for deciding which characters are apomorphic and which are plesiomorphic in a transformation series. The argument goes something like this. As an investigator, you see that members of your group have two different but homologous characters, "round pupils" and "square pupils." As a phylogeneticist, you know that one of these characters, the apomorphic one, might diagnose a monophyletic group, but both cannot (the Grouping Rule). If you think about it, Hennig's reasoning becomes clear. If you find square pupils in the sister group of the taxon you are studying, then it is fairly clear that "square pupils" is older than "round pupils," and if this is true, then "square pupils" must be the plesiomorphic character in the transformation series. Therefore, reasoned Hennig, the characteristics of the sister group are vital in making an intelligent decision regarding polarity within the taxon studied. The simplest rule for determining polarity can be stated in the following way.

Rule for Determining Relative Apomorphy.—Of two or more homologous characters found within a monophyletic group, that character also found in the sister group is the plesiomorphic character, and the one(s) found only in the ingroup is (are) the apomorphic one(s).

As it turns out, actual polarity decisions can be a little more complicated than our simple example. What if, for example, we don't know the exact sister group but only an array of possible sister groups? What if the sister group is a monophyletic group, and it also has both characters? What if our group is not monophyletic? What if "square pupils" evolved in the sister group independently?

POLARITY DECISIONS

The answers to these questions depend on our ability to argue character polarities using some formal rules. The most satisfactory discussion of these rules was published by Maddison et al. (1984). We will present the case developed by them for situations where the phylogenetic relationships among the outgroups and the relationships of these taxa to the ingroup are known. Some handy shortcuts, such as the "doublet rule," will also be covered. However, before we can examine these rules, we need to learn some terms.

The **ingroup node** (IG node) represents the ancestor of the taxon we will eventually be analyzing, once we determine the polarities of our transformation series. The same characters are assigned to the IG node as would be assigned to the ancestral internode, and the terms are interchangeable. If we determine that "square pupils" is primitive, then "square pupils" is assigned to the IG node. To know what is plesiomorphic within the group, we only need to know what character was found at the IG node. In this case, because "square pupils" is assigned to the IG node, the character "round pupils" can be used to diagnose a possible monophyletic group within the IG. Unfortunately, the ancestor isn't around, and thus we must infer what character it had. In Fig. 3.1, the character "square pupils" is coded "a," and "round pupils" is coded "b." The **outgroup node** (OG node) is the node immediately below the IG node. Don't be confused (like some of our students) and think that the OG node refers to characters that are associated with the IG internode; instead, the OG node is associated with characters of the internode immediately below it (in a manner similar to the IG node). So, the general rule is **the internode is associated with the node directly above it.** This is important because characters are usually put on internodes rather than nodes; therefore, it is important to remember that these characters belong to the node *above* them rather than the node *below* them.

Characters are designated by small letters and are placed where taxa are usually labeled. Letters are used purely as a heuristic device and to avoid connotations of primitive and

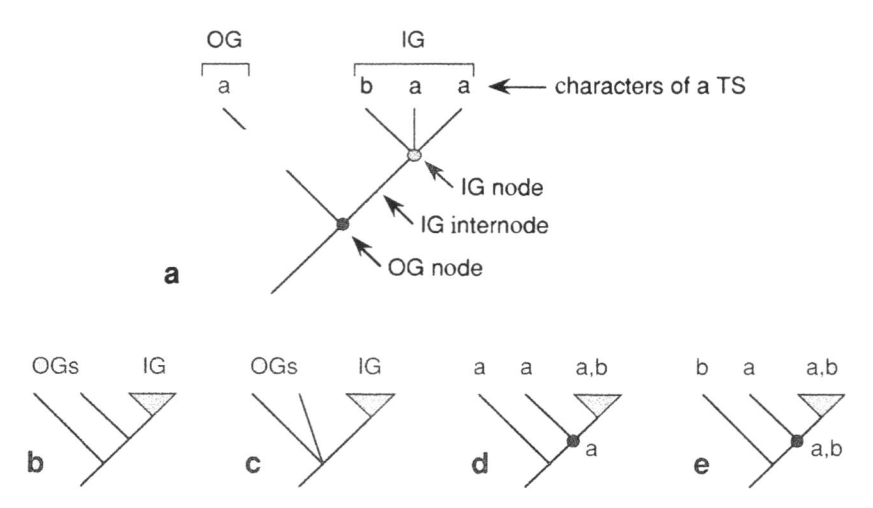

Fig. 3.1.—a. Tree illustrating some general terms used in this chapter. b. Known outgroup relationships. c. Unknown outgroup relationships. d. A decisive character polarity decision with "a" at the OG node. e. An equivocal character polarity decision with "a,b" at the OG node.

CHARACTER ARGUMENTATION AND CODING

derived. The ingroup in Fig. 3.1a is indicated by a polytomy because we presume that the relationships among members are unknown. In all other diagrams, the ingroup is indicated by a shaded triangle, which means exactly the same thing as the polytomy but is easier to draw. Note that the ingroup always has both characters. The relationships among outgroups can either be resolved (Fig. 3.1b) or unresolved (Fig. 3.1c). A decision regarding the character found at the outgroup node may be either **decisive** (Fig. 3.1d) or **equivocal** (Fig. 3.1e). If decisive, then we know that the best estimate of the condition found in the ancestor of our ingroup is the character in question (in this case, "a" is plesiomorphic and "b" is apomorphic). If equivocal, then we are not sure; either "a" or "b" could be plesiomorphic.

Maddison et al. (1984) treat the problem of polarity as one in which the investigator attempts to determine the character to be assigned to the OG node. In effect, it is the character of the ancestor of the ingroup and its sister group (first outgroup) that will give us information about the characters of the common ancestor of the ingroup. Simple parsimony arguments are used in conjunction with an optimization routine developed by Maddison et al. (1984) that was built on the earlier routines of Farris (1970) and Fitch (1971). There are two cases. The first case is relatively complete and is built on known relationships among the outgroups relative to the ingroup. The second case is where the relationships among the outgroups are either unknown or only partly resolved. Because the first case is the simplest, we will use it to describe the general algorithm.

To illustrate the algorithm, we will use the following character matrix (Table 3.1) for the hypothetical group M–S. Sidae is the ingroup, and M, N, O, P, Q, and R are outgroups. The sister group is PQR.

Example 3.1.—Character polarity in the group Sidae.

Table 3.1.—Data matrix for the analysis of Sidae (Example 3.1).

TS	M	N	O	P	Q	R	Sidae
1	b	a	a	b	b		a,b
2	b	b	a	b	b	a	a,b
3	a	b	b	b	b	a	a,b
4	a	a,b	a	b	b	a	a,b

1. Draw the phylogenetic tree of the ingroup and outgroups. You cannot reconstruct the entire tree on the basis of the characters in the matrix shown above. These characters relate to the resolution of relationships in the ingroup, not to the relationships of the ingroup to the outgroup taxa. Presumably, you have either done an analysis or you have used the analysis of another investigator. Figure 3.2 shows the result of this previous analysis with the nodes numbered.

```
        M  N  O  P  Q  R  Sidae
```

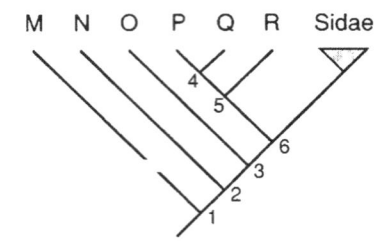

Fig. 3.2.—The relationships of the Sidae and its closest relatives. Outgroups are letters nodes are numbers (Example 3.1).

2. For each transformation series, label each of the branches with the character for that taxon. Use the character matrix for this task. This has been done for TS 1 in Fig. 3.3.

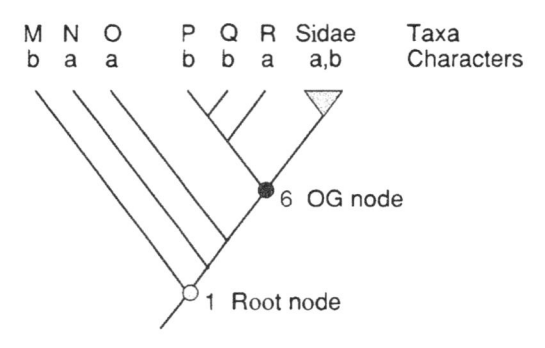

```
    M  N  O   P  Q  R  Sidae    Taxa
    b  a  a   b  b  a  a,b      Characters
```

● 6 OG node

○ 1 Root node

Fig. 3.3.—The relationships of the Sidae and its relatives with characters from TS 1 (Table 3.1) and relevant nodes labeled (Example 3.1).

3. Proceeding from the most distant branches (in this case, M and N), do the following. Label the node "a" if the lower node and adjacent branch are both "a" or "a" and "a,b"; label it "b" if the lower node and adjacent branch are "b" or "b" and "a,b." If these branches/nodes have different labels, one "a" and the other "b," then label the node "a,b." Note that node 1 is not labeled; it is termed the **root node**. For us to label the root, we would need another outgroup. Because we are not interested in the root or in the outgroups, we forget about this node. After all, we are supposed to be solving the relationships of the ingroup. Node 2, the node immediately above the root, is labeled "a,b" in Fig. 3.4 because the first branch (M) is "b" and the second branch (N) is "a."

CHARACTER ARGUMENTATION AND CODING

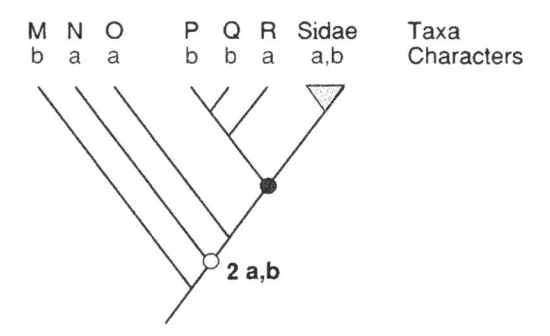

```
M  N  O     P  Q  R  Sidae    Taxa
b  a  a     b  b  a   a,b     Characters
```

Fig. 3.4.—First polarity decision for TS 1, analysis of the Sidae (Example 3.1).

4. Inspect the tree. Do any of the outgroups have a branching structure? In this example, the sister group has a branching structure. For each group of this kind, you need to assign values to their lowest node. So, you work down to the lowest node. Assign to the highest node in such a group a value derived from its two branches. For example, the value "b" is assigned to node 4 in Fig. 3.5.

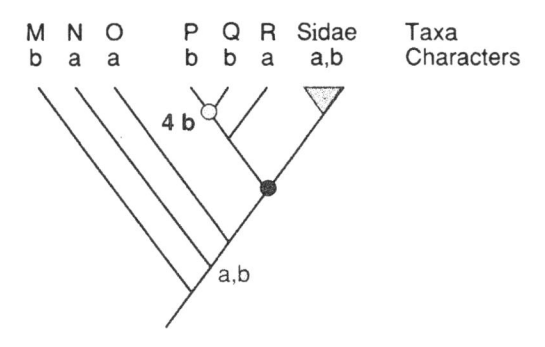

```
M  N  O     P  Q  R  Sidae    Taxa
b  a  a     b  b  a   a,b     Characters
```

Fig. 3.5.—Second polarity decision for TS 1, analysis of the Sidae (Example 3.1).

5. Continue in the direction of the ingroup to the next nodes. For this we use a combination of previous decisions (labeled nodes) and new information from terminal taxa whose ancestral nodes have not been labeled. For example, node 3 in Fig. 3.6 is assigned a decisive "a" based on the "a" of O and the "a,b" of node 2. Node 5 is assigned "a,b" based on the "a" of R and the "b" of node 4.

```
M  N  O     P  Q  R  Sidae     Taxa
b  a  a     b  b  a  a,b       Characters
```

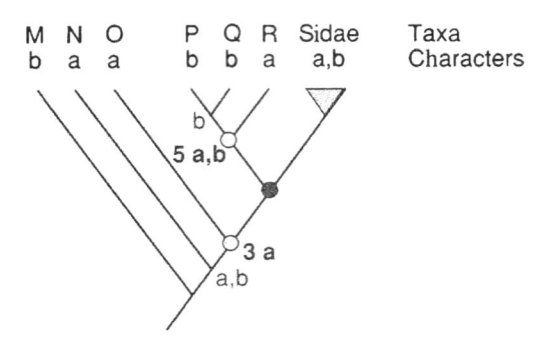

Fig. 3.6.—Third and fourth polarity decisions for TS 1, analysis of the Sidae (Example 3.1).

6. The analysis is over when we reach an assignment concerning the OG node. In this example, the assignment to node 6 is a decisive "a" (Fig. 3.7).

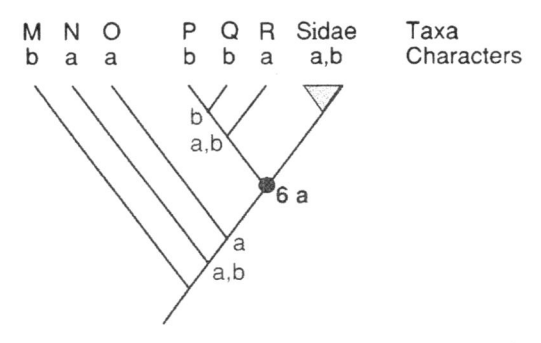

```
M  N  O     P  Q  R  Sidae     Taxa
b  a  a     b  b  a  a,b       Characters
```

Fig. 3.7.—Assignment of polarity to the OG node for TS 1, analysis of the Sidae (Example 3.1).

Figure 3.8 shows characters of TS 2 of the matrix worked out for each node. Note that in this case the decision is equivocal for the OG node.

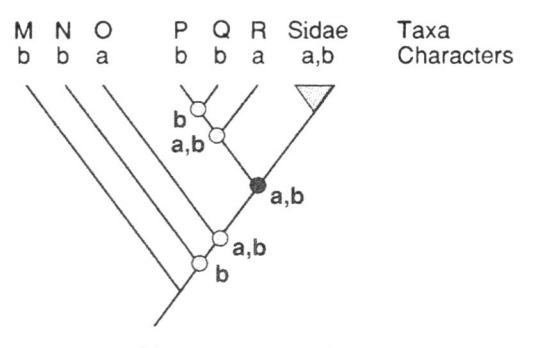

```
M  N  O     P  Q  R  Sidae     Taxa
b  b  a     b  b  a  a,b       Characters
```

Fig. 3.8.—Polarity decisions for TS 2, analysis of the Sidae (Example 3.1).

One last thing. Each of these decisions is made using a single transformation series at a time. This does not mean that equivocal decisions based on single characters taken will remain equivocal at the end of the analysis. The final disposition of character states is subject to overall parsimony rules.

RULES OF THUMB

Maddison et al. (1984) present two rules of analysis that can be used when sister group relationships are known. These rules will help you bypass some of the argumentation for each node of the tree.

Rule 1: The Doublet Rule.—If the sister group and the first two *consecutive* outgroups have the same character, then that character is decisive for the OG node. Any two consecutive outgroups with the same character are called a doublet.

Rule 2: The Alternating Sister Group Rule.—If characters are alternating down the tree, and if the last outgroup has the same character as the sister group, then the character will be decisive for the OG node. If the last outgroup has a different character, then the character decision will be equivocal.

OTHER SITUATIONS

Maddison et al. (1984) also discuss situations in which the relationships among the outgroups are either not resolved or only partly resolved. After you have finished this workbook, you should review their discussion on these topics. We will only mention two important observations. (1) Whatever the resolution of the outgroup relationships, the sister group is always dominant in her influence on the decision. If the sister group is decisive for a particular state, e.g., "a," no topology of outgroups farther down the tree can result in a decisive "b." (2) If you are faced with no sister group but only an unresolved polytomy below the group you are working on, the frequency of a particular character among the outgroups in the polytomy has no effect on the decision for the OG node. For example, you could have 10 possible sister groups with character "a" and one with character "b," and the decision would still be equivocal at the OG node. **Thus, common is not the same as plesiomorphic, even among outgroups.**

Quick Quiz—Outgroups and Polarities

1. Halfway through your phylogenetic study of the saber-toothed cnidaria, your inquiry suffers a fate worse than death. The supposed world's expert, Professor Fenitico, publishes an arrangement lumping your group with its sister group, placing them both in the same genus. How does this affect your analysis?
2. What happens if all the members of the ingroup have a character not found in the sister group or any other outgroup?

Polarity Exercises

For each of the trees and tables below, determine the state to be assigned to the OG node for each of the transformation series. Show the work for each (labeled) node in the outgroup. Prepare a matrix of your decisions using the labeled nodes as taxa. State your decision as equivocal or decisive. The monophyly of the ingroup and the relationships among the ingroup and the outgroups are assumed in each exercise. (No information in the data matrices is relevant to solving the tree shown in each exercise.)

EXERCISE 3.1.—Determine the character assignment for TS 3 and 4 from Table 3.1 for Example 3.1. In TS 4, treat the polymorphic character in taxon N exactly like you would treat an equivocal decision at a node.

EXERCISE 3.2.—Use Table 3.2 and the tree in Fig. 3.9.

Table 3.2.—Data matrix for Exercise 3.2.

Taxon	Transformation series					
	2		3	4		6
A	a		a	a	a	a
B	a	a	a	b	a	b
C	b	a	a	a	b	a
D	b	b	a	b	b	b
E	b	b	b	a	a	b
IG	a,b	a,b	a,b	a,b	a,b	a,b

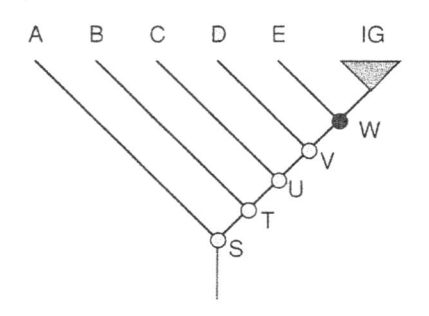

Fig. 3.9.—Tree for Exercise 3.2.

CHARACTER ARGUMENTATION AND CODING

EXERCISE 3.3.—Use Table 3.3 and the tree in Fig. 3.10.

Table 3.3.—Data matrix for Exercise 3.3.

Taxon	\multicolumn{6}{c}{Transformation series}
	2	3	4			6
A	a	b	a	a	a	a
B	a	a	b	a	a	a
C	b	b	a	b	a	b
D	b	b	b	a	b	a
E	b	b	a	a	b	b
IG	a,b	a,b	a,b	a,b	a,b	a,b

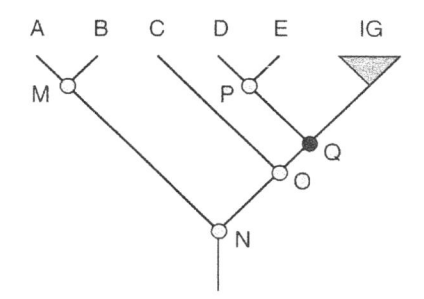

Fig. 3.10.—Tree for Exercise 3.3.

EXERCISE 3.4.—Use Table 3.4 and the tree in Fig. 3.11.

Table 3.4.—Data matrix for Exercise 3.4.

Taxon	\multicolumn{6}{c}{Transformation series}
			3	4		6
A	a	a	a	a	a	a
B	a	b	b	b	b	b
C	a	b	a	b	b	b
D	b	b	a	b	b	b
E	b	b	b	a	a	b
F	a	b	a	b	a	a
G	a	a	a	a	a	a
IG	a,b	a,b	a,b	a,b	a,b	a,b

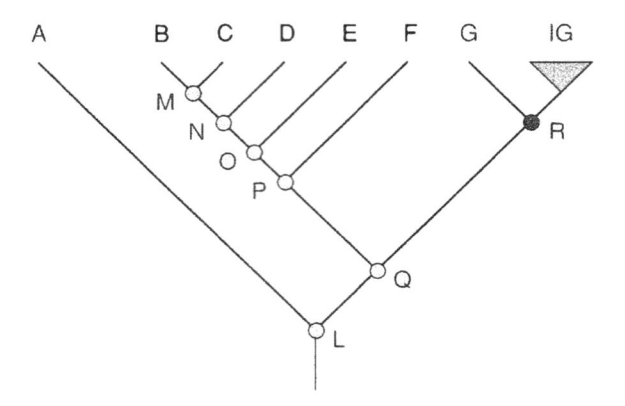

Fig. 3.11.—Tree for Exercise 3.4.

EXERCISE 3.5.—Use Table 3.5 and the tree in Fig. 3.12.

Table 3.5.—Data matrix for Exercise 3.5.

Taxon	\multicolumn{6}{c}{Transformation series}
	2	3	4			6
A	a	a	a	a	a	a
B	a	b	a	b	a	a
C	b	a	a	a	b	b
D	b	a	a	b	b	b
E	a	a	a	b	b	b
F	a	a	b	b	b	b
G	b	b	b	b	b	b
H	b	b	b	a	b	b
I	b	b	a	b	b	b
J	a	b	a	a	a	b
K	a	b	a	b	a	a
IG	a,b	a,b	a,b	a,b	a,b	a,b

CHARACTER CODING

As you learned in Chapter 1, a character is a feature of an organism. A character code is a numerical or alphabetical symbol that represents a particular character. We have already used codes in our previous exercises. By using these characters and their codes, you have learned something about the basics of tree reconstruction using classical Hennig argumentation and some of the approaches to determining the polarity of characters through character argumentation. You already know something about different kinds of characters, homologies, analogies, and homoplasies. In this section, you will be introduced to some of

CHARACTER ARGUMENTATION AND CODING

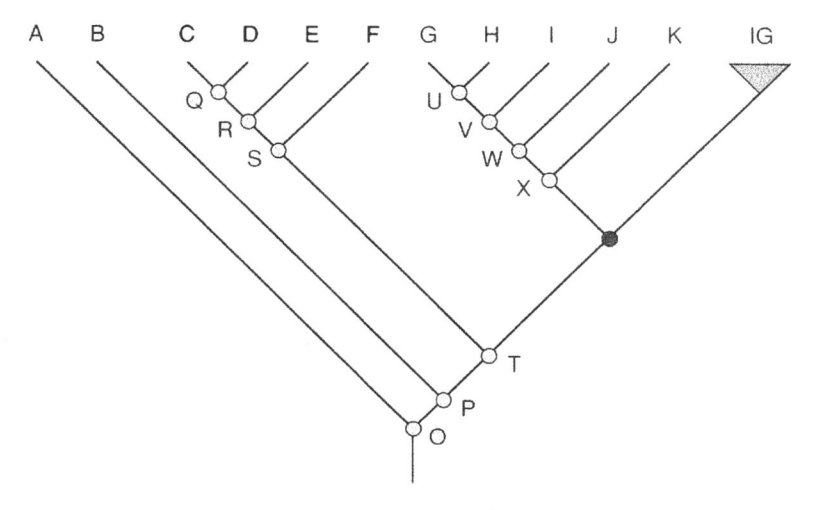

Fig. 3.12.—Tree for Exercise 3.5.

the different kinds of derived characters encountered in phylogenetic research and some of the problems associated with assigning codes to these characters. Before you begin, it might be useful to reread the sections in Chapter 1 about the attributes of specimens. (You should be aware that some investigators refer to transformation series as "characters" and characters as "character states." It is usually quite clear what is being discussed, but this is a potential source of confusion.)

All of the derived characters we have dealt with up to this point are 1) qualitative characters and 2) part of binary transformation series. A binary transformation series consists of a plesiomorphy and its single derived homologue. By convention, the plesiomorphy is coded "0" and the derived homologue is coded "1." As we mentioned in Chapter 1, such binary transformation series are already ordered by virtue of the fact that they are binary. When an investigator works on a large group, or even a small group that has undergone considerable evolution, she may find that there are several different homologous characters in a transformation series. For example, if she were researching the phylogenetic relationships of fossil and Recent horses, the transformation series containing the characters for the number of toes of the hind foot would contain four different but related characters: four toes, three toes, and one toe in the ingroup and five toes in the outgroups. This kind of transformation series is termed a **multistate transformation series**. A multistate transformation series contains a plesiomorphic character and two or more apomorphic characters.

Simple binary transformation series present no problem in coding. The investigator codes by outgroup argumentation, according to the information available, producing a matrix full of 0 and 1 values. You have practiced this kind of coding in Chapter 2. Complications arise if there are one or more **polymorphic taxa**, i.e., taxa with both the plesiomorphic and apomorphic characters. The problem is only critical when both types of characters are found in a single species. In such cases, the taxon can be treated as having both characters, a designation easily handled by available computer programs.

Multistate transformation series can be grouped in two ways: 1) according to what we know of their evolution and 2) according to the way they are related. **Ordered transformation series** are those in which the relationship of characters within the transformation series are specified (and presumably "known"). Binary transformation series are the simplest case of ordered transformation series. A **polarized ordered transformation series** is one in which not only the relationships are specified, but also the direction of evolution. **Unordered transformation series** are those where the relationships of characters to one another are not specified. This does not mean that we know nothing about the transformation series. Frequently, we know which of the characters is plesiomorphic; we just don't know the order of the derived transformation series. Such a transformation series is partly polarized. If you are using a computer program such as PAUP or MacClade, the program can tell which of the unordered characters is the plesiomorphic one when you root the tree at a specific point or with a specific taxon. The program relies on you, however, to specify this information correctly.

The second way of grouping multistate transformation series is by their relationships. Polarized transformation series may come in several varieties. In the simplest case, the characters might be related in a linear fashion. A **linear transformation series** consists of characters related to one another in a straight-line fashion such that there are no branches on the character tree (Fig. 3.13). The relationships of these characters can be termed a **character tree**. It is important to understand that a character phylogeny is not the same as a phylogeny of taxa. A character tree contains only information about the relationships among characters; the distribution of these characters among taxa is shown in descriptions diagnoses, and character matrices.

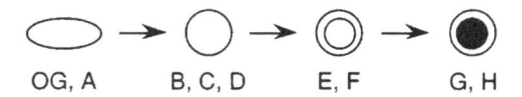

Fig. 3.13.—A simple linear character tree of four characters. Letters represent taxa in which each character is found.

Linear transformation series present no problems in coding; one simply assigns a value to each character in ascending order. Each value is placed in the data matrix in a single column, and each apomorphy contributes to the length of the tree in an additive fashion. We use the term additive because each instance of evolution is one step along the tree, and counting all of the steps in a straight line shows exactly how much the transformation series has added to the overall tree length. (Such transformation series are often termed additive multistate characters.)

CHARACTER ARGUMENTATION AND CODING   37

A **branching transformation series** contains characters that are not related to each other in a straight-line fashion (Fig. 3.14). Such transformation series may present problems because the relationships among the characters are represented by a branching pattern rather than a straight-line pattern. Because of this, the characters cannot be coded in an additive fashion. Such transformation series are also called nonadditive or complex transformation series. Because the characters are not related in a linear fashion, simple additive coding will result in errors in translating the transformation series into a phylogenetic tree.

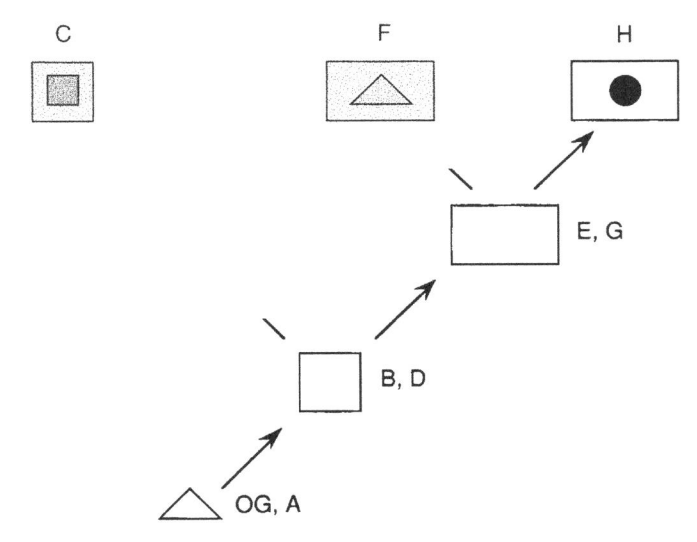

Fig. 3.14.—A complex branching character tree of six characters. Letters represent taxa in which each character is found.

We present two examples of character coding using three techniques. You will probably ask where we came up with the character trees of the two transformation series. This is a good question and one we will return to in a later section. For now, we are only concerned with the formalities of coding and not how one actually determines the character trees.

Example 3.2.—A simple linear transformation series.

In Fig. 3.13, we show a simple linear transformation series of four characters. Below the character tree is an account of the distribution of these characters among nine taxa. The data matrix for this transformation series can be constructed in one of two basic ways. First, we can simply code the transformation series in a linear fashion, assigning a value to each character based on its place in the character phylogeny. We have chosen to code with values ranging from 0 to 3 (Table 3.6).

Table 3.6.—Data matrix for a simple linear transformation series coded by the linear and the additive binary methods (Example 3.2).

Taxon	Linear coding	Additive binary coding*		
		C + C/C + C/D	C/C + C/D	C/D
OG	0	0	0	0
A	0			
B	1			
C	1			
D	1			
E	2			
F	2			
G	3			
H	3			

* Column heads are apomorphic (1) characters. C = circle; C/C = circle/circle; C/D = circle/dot.

We could also use **additive binary coding**, which is a method that breaks the character down into a number of subcharacters, each represented by its own column of information. For example, because the characters in the transformation series really consist of subsets of related characters, we can consider both "circle/circle" and "circle/dot" as subsets of "circle" because each is derived from "circle" later in the character tree. The first additive binary column in Table 3.6 reflects this fact, coding "oval" as the plesiomorphic character (0) and "circle" plus all of its descendants as apomorphic (1). "Circle/dot" is a subset of "circle/circle." Both "oval" and "circle" are plesiomorphic relative to "circle/circle" so they get a coding of "0," whereas "circle/circle" and its descendant "circle/dot" get a coding of "1." Finally, "circle/dot" is apomorphic relative to "circle/circle" (and "oval" and "circle") so it is coded "1" in the third column. So in total, we have produced three columns to represent the transformation series. Now, go along the rows and add up all of the 1's in the additive binary matrix and put them into a single column. You should find that you have replicated the original linear transformation series of 0-1-2-3. Either method of coding produces exactly the same phylogeny. But, there are some differences. If you use binary coding, you must keep in mind that formal computer algorithms and the programs that use them cannot tell the difference between three noncorrelated and independent transformation series and a single binary coded transformation series. This doesn't cause problems with phylogenetic analysis, but it can produce results that seem strange in biogeographic analysis and the analysis of coevolutionary patterns, which we will discuss in Chapter 7

Example 3.3.—A branching transformation series.

Let us look at the branching transformation series in Fig. 3.14. The taxa sharing a particular character are shown beside or above the character on the character tree. This transformation series is considerably more complex than the first one. It should be obvious that a single labeling of characters in a linear fashion would result in some misinformation. How do we show these complex relationships? There are two basic methods, nonadditive

CHARACTER ARGUMENTATION AND CODING

binary coding and mixed coding. Because we have already seen an example of binary coding, let us turn to this method first.

Review the character tree and then examine the nonadditive binary codings in Table 3.7. Note that "square" is apomorphic relative to "triangle," by outgroup comparison, and that "square" is ancestral to all other characters in the character tree. Our first binary column reflects this fact: "square" and all of its descendants are coded "1," whereas "triangle" is coded "0." "Square/square" is derived from "square." "Rectangle" is also directly derived from "square." Look at "square/square." It is only found in taxon C. We produce a new column reflecting this fact. In this column "square/square" acts like an autapomorphy (which it is). "Rectangle" does not act as an autapomorphy; it is plesiomorphic to two other characters. This fact is used to code "rectangle" in a similar manner to the way we coded "square," as shown in the third column. Since both "rectangle/triangle" and "rectangle/dot" are unique to their respective taxa, F and H, we code them in a manner similar to "square/square." Now, you are able to reconstruct the phylogeny of the group AH using the two character phylogenies.

Table 3.7.—Data matrix for a branching transformation series coded by the nonadditive binary and the mixed methods (Example 3.3).

Taxon	Nonadditive binary coding*					Mixed coding*		
	All except T	S/S	R+	R/D	R/T	T+S+R+R/D	S/S	R/T
OG	0	0	0	0	0	0	0	0
A	0	0	0	0	0	0	0	0
B	1	0	0	0	0	1	0	0
C	1	1	0	0	0	1	1	0
D	1	0	0	0	0	1	0	0
E	1	0	1	0	0	2	0	0
F	1	0	1	1	0	2	0	1
G	1	0	1	0	0	2	0	0
H	1	0	1	0	1	3	0	0

* Column heads are apomorphic (≥1) characters. T = triangle; S = square; S/S = square/square; R+ = rectangle and all descendants; R/D = rectangle/dot; R/T = rectangle/triangle.

Mixed coding is a hybrid between additive binary coding and linear coding. Mixed coding has also been termed nonredundant linear coding. By convention, the longest straight-line branch of the character tree is coded in a linear fashion. Branches off this linear tree are coded in an additive binary fashion. This strategy might save character columns, depending on the asymmetry of the character tree. We can code the section of the character tree that goes "triangle," "square," "rectangle," "rectangle/dot" in a single column (0-1-2-3) in the first column of Mixed coding in Table 3.7. (How do you know to use "rectangle/dot" as the fourth character in the transformation series? Actually, the choice is completely arbitrary; remember, nodes can be freely rotated. We could have just as well used "rectangle/triangle" and coded "rectangle/dot" as the autapomorphy.) A separate column is then used for "square/square" (column 2), and a final column for "rectangle/triangle."

Basal bifurcations occur when both the outgroup and one of the ingroup taxa have the same character. In this case, assign "1" rather than "0" to the most plesiomorphic character and proceed. This coding strategy serves to "link" the columns and will not add steps to the tree.

Note on character coding. Newer computer algorithms such as PAUP 3.0 can use a character tree directly if the investigator inputs the relationships of the characters. It then uses this information to construct an additive binary matrix for analysis, which the investigator never sees (Swofford, 1990).

Quick Quiz—Character Coding

1. Your research into the systematics of the spade-lipped mugmorts is halted until you resolve the coding of a troublesome transformation series. You have identified the plesiomorphic character, but the remaining six cannot be polarized. What type of transformation series should be considered?
2. The copulatory organ of spade-lipped mugmorts has various colors, including no color at all. The evolutionary sequence of color change is not known except that the sister group and all other outgroups have colorless organs. Although you could opt for a polarized but unordered transformation series, you opt instead for a binary transformation series for each color (e.g., no color [0] to blue [1], no color [0] to green [1], etc.). What effect will your decision have on reconstructing the tree?

Coding Exercises

For each of the trees shown below do the following:

1. Determine the possible types of coding strategies that might be used and list them.
2. Explain why certain coding strategies cannot be employed for the particular character tree.
3. Prepare a data matrix for each type of coding strategy you think could be employed.
4. Solve the phylogenetic problem with the data in the matrix.

EXERCISE 3.6.—Use Fig. 3.15.

l ⟶ m ⟶ n ⟶ o
OG   A   B   C

Fig. 3.15.—A character tree for four characters. Capital letters represent taxa in which each character (lowercase letters) is found. We use letters rather than numbers to emphasize the difference between a character and a character coding.

CHARACTER ARGUMENTATION AND CODING

EXERCISE 3.7.—Use the character tree in Fig. 3.16.

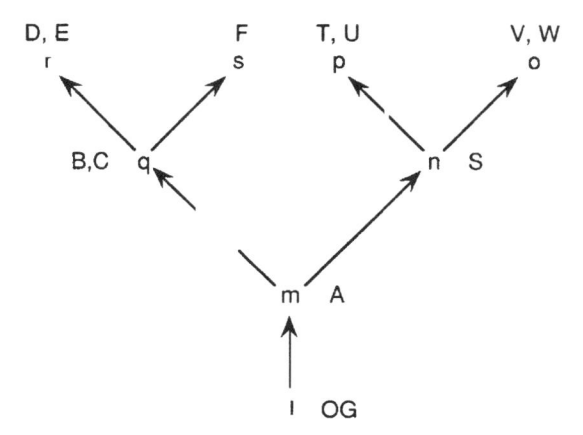

Fig. 3.16.—A character tree for eight characters. Capital letters represent taxa in which each character (lowercase letters) is found.

EXERCISE 3.8.—Use the character tree in Fig. 3.17.

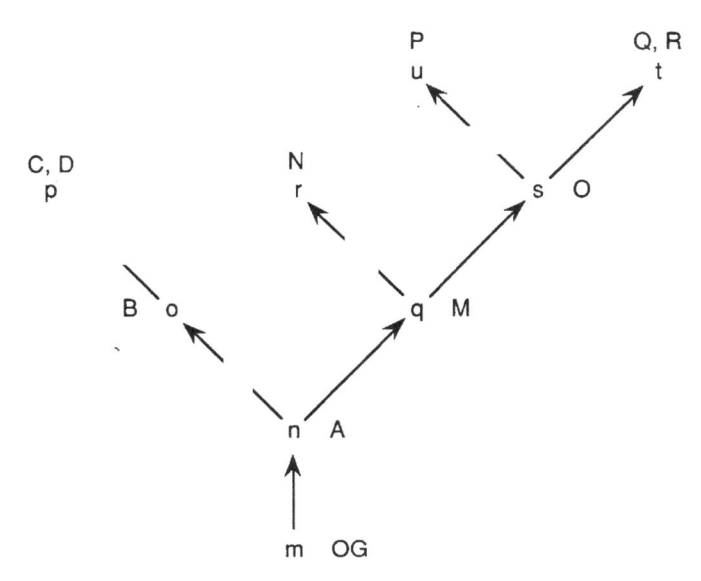

Fig. 3.17.—A character tree for nine characters. Capital letters represent taxa in which each character (lowercase letters) is found.

Chapter Notes and References

1. Discussions of homology, different kinds of characters, and basic character argumentation can be found in Wiley (1981a). However, this reference is outdated when it comes to outgroup comparison and contains no information of use on character coding and other "modern" issues.

2. There is a lot of information on outgroups and outgroup comparisons. Maddison et al. (1984) was preceded by Watrous and Wheeler (1981) and a criticism of them by Farris (1982). Wiley (1987b) contains a summary of the three papers. The discussion by Crisci and Stuessy (1980) is, in our opinion, positively misleading and should be avoided. Donoghue and Cantino (1980) discuss one method of outgroup comparison, the outgroup substitution method, that can be useful when relationships among outgroups are problematic.

3. Those who have read some phylogenetic literature will note that we have avoided, until now, any mention of other criteria. Wiley (1981a) discusses several other criteria. Other useful discussions can be found in de Jong (1980) and Stevens (1980). The major bone of contention is what is known as the ontogenetic criterion. Some, such as Nelson (1978, 1985), Patterson (1982), Rieppel (1985), and Weston (1988), advocate the ontogenetic criterion as a (or *the*) major criterion for determining polarity. We do not think this is a general criterion (see Brooks and Wiley, 1985; Kluge, 1985, 1988a; O'Grady, 1985; Kluge and Strauss, 1986), but we recognize that it can be used to both check hypotheses of homology (cf. Hennig, 1966; Wiley, 1981a; Patterson, 1982; Kluge, 1988a) and infer polarity under certain assumptions. Before you employ this criterion, you should read Mabee (1989).

4. Some papers of interest on character coding include Farris et al. (1970), Mickevich (1982), O'Grady and Deets (1987), Pimentel and Riggins (1987), and O'Grady et al. (1989).

Complications arise if one or more of the taxa have both the plesiomorphic and apomorphic character. Polymorphic taxa have both the plesiomorphic and apomorphic characters. Actually, the problem is only critical when both are found in a single species. Considerable controversy surrounds the coding of such characters, especially when biochemical characters are used. There are two ways of handling such characters: 1) coding the taxon as having the apomorphy only and discounting the plesiomorphy or coding both characters as present and using a computer program such as PAUP that can handle polymorphic data cells (qualitative coding) or, 2) coding according to frequency of each character. Swofford and Berlocher (1987) present a strong case for analysis of frequency data and suggest computational methods for accomplishing this within a phylogenetic analysis. D. L. Swofford (pers. comm.) has authored a computer program (FREQPARS) to accomplish this task. Buth (1984) is an excellent introduction to the use of electrophoretic characters.

Quick Quiz Answers

Outgroups and Polarities

1. This question has no simple answer. We suggest the following. Examine the paper. Has Professor Fenitico provided synapomorphies to support his argument? If not, then he has only produced another arrangement and not a scientific hypothesis you can evaluate, so you should proceed with your problem as if nothing had been published. If he does provide synapomorphies, what is the nature of these characters? Do they demonstrate that the sister group is monophyletic? Is the sister group still the sister group, even if it is now in the same genus? If so, then the nature of our character argumentation has not changed, only the taxonomy, which might be very important to Professor Fenitico but should not be important to you. However, if Professor Fenitico has demonstrated that the supposed sister group is really embedded within your group, then take this into consideration, redesign your arguments, and write Professor Fenitico to tell him that names really don't mean anything, especially his.

2. If this character is really unique to the ingroup, then it is a synapomorphy of the members of the group (or, if you wish, an autapomorphy of the group).

Character Coding

1. You can opt for an unordered transformation series or you can try coding six binary ordered series. If you pick the binary series, check answer 2.

2. Because "no color" is symplesiomorphic, repeated use of this character in different transformation series will result in an answer that has no bearing on the relationships among the taxa. The effect is to render all of the color characters autapomorphic, which implies that all are independently derived from "no color." Better see answer 1 and opt for an unordered transformation series.

CHAPTER 4

TREE BUILDING AND OPTIMIZATION

Phylogeneticists frequently describe their work in terms of "building trees" or "reconstructing phylogenies." These activities are directed towards attempts to discover something we believe exists in nature, the common ancestry relationships among the organisms we study. Interestingly, modern computer programs do not spend much computing time building trees. Rather, most of the time is spent evaluating different tree topologies (branching patterns) in an effort to find the tree that meets a criterion of optimality given the data. How the tree is actually generated may be irrelevant. For example, you can evaluate all of the possible trees for a three taxon problem by simply mapping the character distributions on the four possible trees in the most efficient manner (i.e., maximizing the number of synapomorphies and minimizing the number of homoplasies needed given the tree). You don't have to build a tree, all the possible trees are given. Under the criterion that the shortest tree is the optimal tree, all you have to do is count the changes and pick the shortest tree among the four possibilities.

Of course, as the number of taxa increases, the number of possible trees increases very quickly (see Chapter 6). Because phylogenetic methods were originally built around constructing trees, many of the classic works emphasize reconstruction, and they do so using methods such as Hennig argumentation (Hennig, 1966) and the Wagner algorithm (Kluge and Farris, 1969). Although many "modern" phylogeneticists will never use classic Hennig argumentation and algorithms such as the Wagner algorithm provide only the starting point for some (not even all) modern computer programs, it is important for you to get the feel of these approaches because they give insight into the nature of phylogenetic data and help you understand how previous investigators arrived at their conclusions. Thus, we have organized this chapter in a quasi-historical fashion. We begin with Hennig argumentation, Hennig's own method for reconstructing phylogenies. We then use the Wagner algorithm to teach the rudiments of a formal algorithm and how such an algorithm might be implemented and provide some basic terms encountered in more modern methods. We then discuss the concepts of the optimal tree, optimal character distribution, various parsimony criteria, and ACCTRAN and DELTRAN optimization. Finally, we provide a very brief discussion of how current algorithms operate to produce optimal or near optimal trees.

HENNIG ARGUMENTATION

You already have had practice at performing analyses using Hennig argumentation in Chapter 2. However, you did it in a rather laborious way, using the inclusion/exclusion principle. Hennig argumentation was the original phylogenetic algorithm, and its application is still common. For simple problems, Hennig argumentation presents no technical problems. However, even with relatively few taxa and characters, you will find it much too tedious to make all of those single-character trees and all of the logically incompatible alternative trees. In computer implementations of phylogenetics, the computer performs this boring task. An investigator working without a computer has two problems. First, she would never want to make all the inclusion/exclusion trees. Second, with even a small amount of homoplasy the investigator runs the chance of missing equally parsimonious solutions to the tree that she finds, not because the algorithm is defective but because the human mind rejects the ennui of considering all possible alternative trees. A really experienced phylogeneticist instead will "inspect" the data matrix and produce a first tree based on this general inspection, filtering through the data in her mind. We would like you to try this on the exercises below. One strategy, for example, would be to circle potential groupings within the data matrix. Another strategy is to start with "obvious groups" (=lots of synapomorphies) and then attempt to link them together. Although this might seem rather imprecise to you, remember that it is similar to the method Hennig himself probably used. Remember, no tree you draw has to be the final tree. **All trees are hypotheses.**

Hennig Exercises

EXERCISE 4.1.—Bremer's (1978) *Leysera* data.

Leysera is a small genus of composite shrublets found in southern Africa (three species) and the Mediterranean region (one species). The closest relatives of *Leysera* appear to be the genera *Athrixia, Rosenia,* and *Relhandia.* The phylogenetic relationships of these genera are as follows. *Leysera, Rosenia,* and *Relhandia* form a trichotomy. *Athrixia* is the sister of these three genera. *Leysera* is monophyletic based on two characters: 1) the chromosome number is $2N = 8$ and 2) all have a solitary capitula on a long peduncle. The distribution of characters among the four species of *Leysera* is given in Table 4.1.

Table 4.1.—*Leysera* ingroup characters.

Taxon	Receptacle	Floret tubules	Pappus type	Achene surface	Pappus scales	Life cycle
L. longipes	smooth	with glands	barbellate	smooth	subulate	perennial
L. leyseroides	rough	with hairs	plumose	rough	wide, flat	annual
L. tennella	rough	with hairs	plumose	rough	wide, flat	annual
L. gnaphalodes	rough	with hairs	plumose	rough	subulate	perennial

Based on outgroup information, the following characters are plesiomorphic: 1) receptacles smooth, 2) hairs absent on the floret tubules, 3) barbellate pappus, 4) achene surface smooth, 5) pappus scales subulate, and 6) perennial life cycle.

1. Prepare a data matrix.
2. Analyze the phylogenetic relationships of *Leysera* based on the information given and draw the tree of relationships.

EXERCISE 4.2.—Siegel-Causey's (1988) cliff shags.

Shags, cormorants, and anhingas comprise a clade of marine and littoral fish-eating birds. Among shags, the genus *Stictocarbo* (cliff shags) comprises eight species. In this exercise, you will use both outgroup information and ingroup information to reconstruct the relationships among six species of Stictocarbo. The seventh species (*S. magellanicus*) is used as the sister group, and species in other genera provide additional outgroups.

1. Using the tree in Fig. 4.1 and the characters in Table 4.2, determine the character at the outgroup node for each transformation series and arrange this as a character vector labled "OG."

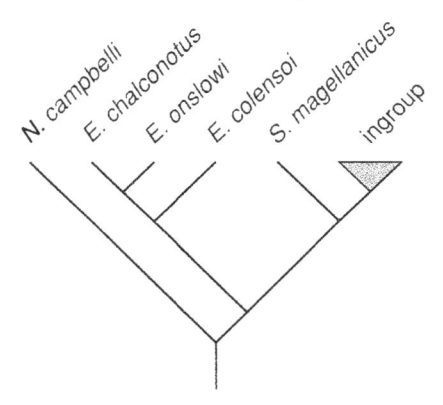

Fig. 4.1.—A hypothesis of the phylogenetic relationships among certain shags.

2. Adding the characters in Table 4.3, reconstruct the phylogenetic relationships among the remaining six species of the genus. **Tip:** Some of the decisions you reached in step 1 are equivocal. Don't use these transformation series to reconstruct the initial tree.

3. After reconstructing the relationships among the species, examine the transformation series with equivocal decisions at the outgroup node. Can you now characterize them? Will one or more remain equivocal, giving rise to alternative interpretations of character evolution?

THE WAGNER ALGORITHM

The basics of the Wagner algorithm were published by Kluge and Farris (1969). Although phylogenetic, the algorithm was developed independent of the Hennig argumentation algorithm and was based on Wagner Groundplan Divergence Analysis.

Table 4.2.—Shag outgroup characters for use in Exercise 4.2. Numbers for transformation series correspond to those found in Siegel-Causey (1988).

Taxon	Transformation series																				
	1	2	22	36	39	40	42	48	63	69	78	79	81	94	97	100	102	110	111	112	114	120	124	131	134
N. bulleri	0	0	0	0	0	0	0	0	0	0	0	1	0	0	0	0	0	0	0	0	0	0	0	0	0
E. chalconotus					0	1									0	1									0
E. ? ×ef					00	1									0	1									0
E. colensoi					0	1									0	1									0
S. magellanicus	0	1	0	1	0	1	0	0	0	0	0	1	0	0	0	1	1	0	0	0	0	0	1	0	0

Table 4.3.—Shag ingroup characters for use in Exercise 4.2. Numbers for transformation series correspond to those found in Siegel-affy (1988).

Taxon	Transformation series																				
	1	2	22	36	39	40	42	48	63	69	78	79	81	94	97	100	102	110	111	112	114	120	124	131	134
S. pelagicus	0	1	0	0	0	0	1	0	0	0	0	0	0	0	0	1	1	0	0	0	0	0	0	0	1
S. urile	0	1	0	0	0	1	0	0	0	1	0	0	0	0	1	1	0	1	0	0	0	1	1	1	0
S. ×tis	0	0	1	0	0	1	0	0	1	0	0	0	0	0	1	0	1	1	1	0	0	0	0	0	0
S. gaimardi			0	1		0	0	0	1	1	0		0	1	1	1	1	1	1	1	1	0	0	0	0
S. ×tus	1	0	0	0	1	0	0	0	1	0	1	0	0	1	1	1	0	1	1	1	0	0	1	0	0
S. featherstoni	1	0	0	0	0	1	0	0	0	1	0	0	1	1	1	1	0	1	1	0	0	0	0	0	0

Wagner Definitions

Like all techniques, the Wagner algorithm comes with certain terms that must be learned to appreciate its logic. In some cases, these terms are illustrated by examples (refer to Table 4.4).

Table 4.4.—Sample data matrix for Wagner definitions and Exercise 4.3.

Taxon	Characters				
M	1	0	0	0	0
A	0	1	0	1	1
B	0	1	1	0	0

1. A particular **character** (X) of a particular **taxon** (A) is defined as $X(A,i)$, where i is the ith character in a vector of i characters.

2. The **vector** of characters for a particular taxon is defined as $\Sigma X(A,i)$. For example, the character vector for M is

$$\Sigma X(M,i) = 1\ 0\ 0\ 0\ 0.$$

3. The **difference** (D) between two taxa is the sum of the absolute differences between their characters:

$$D(A,M) = \Sigma |X(A,i) - X(M,i)|.$$

We calculate this in the following manner:

$$\begin{aligned} D(A,M) &= \Sigma |X(A,i) - X(M,i)| \\ &= |0-1| + |1-0| + |0-0| + |1-0| + |1-0| \\ &= 4. \end{aligned}$$

EXERCISE 4.3.—Calculate the difference (D) between A and B and between M and B (Table 4.4).

4. The **interval** (INT) of a taxon is the length of the line between that taxon and its ancestor. For example, the interval of B is

$$INT(B) = D[B,ANC(B)],$$

where INT(B) is the interval of taxon B, ANC(B) is the hypothetical ancestor of B, and D[B,ANC(B)] is the path length distance of B to its ancestor.

Example 4.1.—Calculating interval B.

Let us take our simple data matrix and calculate an interval. Designate M as the ancestor (it's really the outgroup, but it doesn't matter here). So, ANC(B) is M and the formula reads

$$\text{INT(B)} = D[B,M]$$
$$= \sum |X(B,i) - X(M,i)|$$
$$= |0 - 1| + |1 - 0| + |1 - 0| + |0 - 0| + |0 - 0|$$
$$= 3.$$

The interval is shown graphically in Fig. 4.2.

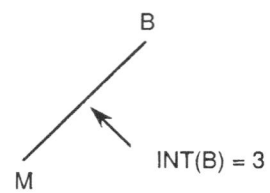

Fig. 4.2.—Graphic representation of INT(B) of the Wagner algorithm.

EXERCISE 4.4.—Distances and intervals (Table 4.5).

Table 4.5.—Data matrix for Wagner algorithm distances and intervals (Exercise 4.4).

Taxon	Characters					
ANC	0	0	0	0	0	0
A	1	1	0	0	0	0
B	1	0	1	0	1	0
C	0	0	1	1	0	1

For this exercise we will make the following assumptions. First, all of the characters scored "1" have been previously argued as derived, using the standard outgroup procedures. Second, we have used outgroup comparison to construct a valid hypothetical ancestor. (Note that the ancestral vector is composed entirely of "0" values.) From the data in Table 4.5, 1) calculate the distance of each taxon to each other taxon and from each taxon to the ancestor and 2) pick the least distance and calculate the interval from that taxon to the ancestor.

The Algorithm

Given any matrix of characters, we implement the Wagner algorithm in the following manner (from Kluge and Farris, 1969).

1. Specify an ancestor or outgroup.
2. Within the ingroup, find the taxon that shows the least amount of difference from the ancestor/outgroup. To accomplish this, calculate D for each taxon to the ancestor/outgroup.
3. Create an interval for the taxon that has the smallest D.

4. Find the next taxon that has the next smallest difference from the ancestor/sister group. Do this by inspecting the original D values you calculated. If there is a tie (i.e., two or more with the same value of D), then arbitrarily pick one.

5. Find the interval that has the smallest difference with the taxon selected in step 4. To compute D(taxon,interval), use the following formula:

$$D[B,INT(A)] = \frac{D(B,A) + D[B,ANC(A)] - D[A,ANC(A)]}{2}.$$

6. Attach the taxon to the interval you have selected by constructing a hypothetical ancestor for the two taxa. The character vector of the ancestor, and thus its position along the interval, is computed by taking the median value of the existing taxon, its ancestor, and the added taxon.

7. Go to step 4 and repeat for each remaining taxon.

Example 4.2.—First Wagner calculations.

We begin with an extremely simple example taken from Wiley (1981a). There are three taxa and their ancestor. Step 1 is specified in Table 4.6 as the ANC vector.

Table 4.6.—Data matrix for Wagner calculations (Example 4.2).

Taxon	Characters					
ANC	0	0	0	0	0	0
A	1	1	0	0	0	0
B	1	0	1	0	1	0
C	1	0	1	1	0	1

2. Calculate D from the ancestor for each taxon. You already did this in Exercise 4.5.

$$D(A,ANC) = 2$$
$$D(B,ANC) = 3$$
$$D(C,ANC) = 4$$

3. Construct an interval for this taxon. A has the smallest distance to ANC, so we construct INT(A,ANC):

$$INT(A,ANC) = D(A,ANC) = 2.$$

4. Select the next taxon that has the smallest D to the ANC. This would be taxon B.

5. Find the interval that has the smallest D to taxon B. Because there is only one interval, INT(A,ANC), we have no choice but to add B to this interval. Therefore, we don't have to compute D[B,INT(A)]. We connect B to INT(A) by constructing a hypothetical ancestor (X) whose characters are the median of the transformation series of ANC, A, and B, the three taxa involved in the problem at this point (Table 4.7). Our tree now has a branch, a new

hypothetical ancestor, and, most importantly, three intervals, INT(A), INT(B), and INT(X) (Fig. 4.3).

Table 4.7.—Data matrix for Wagner calculations with a hypothetical ancestor (X) (Example 4.2).

Taxon	Characters					
ANC	0	0	0	0	0	0
A	1	1	0	0	0	0
B	1	0	1	0	1	0
X (median)	1	0	0	0	0	0

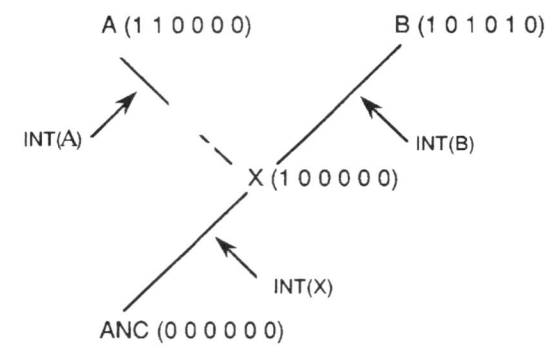

```
         A (1 1 0 0 0 0)        B (1 0 1 0 1 0)
                  ↖             ↗
         INT(A)      ↘         ↙     INT(B)
                       X (1 0 0 0 0 0)
                           ↖
                              INT(X)
                    ANC (0 0 0 0 0 0)
```

Fig. 4.3.—A branched Wagner tree with a new hypothetical ancestor (X) and three intervals (Example 4 2).

6. We return to step 4 of the algorithm, adding the next taxon that shows the least difference from ANC. By default, this is taxon C, but where do we add C? The algorithm states that it should be added to that interval that has the smallest difference from C. Therefore, we must calculate three interval difference values, one for each interval in the tree. The formula for figuring the difference between a taxon and an interval requires finding the difference between the taxon added and the ancestor of the taxon already in the tree. In this case, there are two ancestors. You already know the difference between C and ANC (D = 4, see above). But, you haven't calculated the difference between A, B, or C and the new ancestor, X, and you haven't calculated the differences between C and A or C and B. This is the first step.

$$D(A,X) = |X(A,i) - X(X,i)| = 1$$
$$D(B,X) = |X(B,i) - X(X,i)| = 2$$
$$D(C,X) = |X(C,i) - X(X,i)| = 3$$
$$D(C,A) = |X(C,i) - X(A,i)| = 4$$
$$D(C,B) = |X(C,i) - X(B,i)| = 3$$
$$D(X,ANC) = |X(C,i) - X(ANC,i)| = 1$$

Now we can begin our calculations.

$$D[C, INT(A)] = \frac{D(C,A) + D(C,X) - D(A,X)}{2}$$

$$= \frac{(4 + 3 - 1)}{2}$$

$$= 3$$

$$D[C, INT(B)] = \frac{D(C,B) + D(C,X) - D(B,X)}{2}$$

$$= \frac{(3 + 3 - 2)}{2}$$

$$= 2$$

$$D[C, INT(X)] = \frac{D(C,X) + D(C,ANC) - D(X,ANC)}{2}$$

$$= \frac{(3 + 4 - 1)}{2}$$

$$= 3$$

Because the difference between C and INT(B) has the smallest value, we construct another hypothetical ancestor (Y) and connect C to the tree through this new ancestor to INT(B) (Fig. 4.4). To calculate the character vector for this new ancestor, take the median of the vectors of the three appropriate taxa, X, B, and C (Table 4.8). You have now completed the problem (Fig. 4.4).

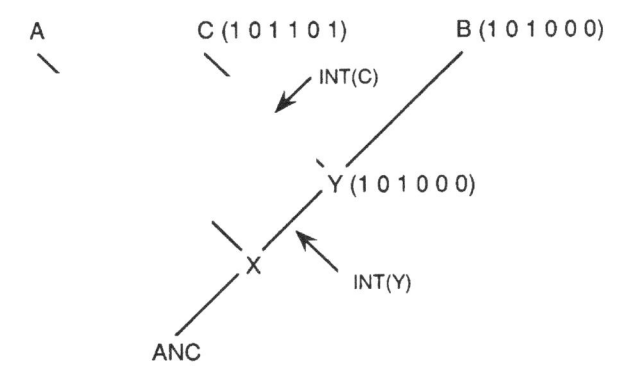

Fig. 4.4.—Complete Wagner tree with two hypothetical ancestors (X and Y) (Example 4 2).

Table 4.8.—Complete data matrix for Wagner calculations with a second hypothetical ancestor (Y) (Example 4.2).

Taxon	Characters					
X		0	0	0	0	0
B		0	1	0	1	0
C		0	1	1	0	1
Y (median)	1	0	1	0	0	0

Wagner Tree Exercises

The two exercises use the same data matrices as those used in the Hennig argumentation exercises. To follow the exercise answers more closely, we suggest that you label the ancestors sequentially beginning with "A" and follow the tips in each exercise.

EXERCISE 4.5.—Bremer's (1978) *Leysera* data.

Reconstruct the relationships among Bremer's species (Table 4.1) using the Wagner algorithm. Use a zero vector ancestor. **Tip:** There is a tie; add *L. leyseroides* before *L. tennella* if you want to conform directly with the answers.

EXERCISE 4.6.—Siegel-Causey's (1988) cliff shags.

Reconstruct the relationships of Siegel-Causey's cliff shags using Table 4.3 and the character vector you reconstructed at the OG node. Some of the character decisions at the OG node were equivocal; for this exercise use the following decisions for these transformation series: 2-0, 36-1, 102-0, 124-1. This is a large and complicated exercise, but it will be worth the effort to complete. **Tips:** 1) To decrease the work load, compare ancestral intervals before you compare terminal intervals. If the distance to the interval of, say, ancestor B is greater than that to ancestor C, then you will not have to compute the distances to those terminal intervals connecting to ancestor B. This shortcut works because we are using path length distances. 2) You will find two ties: add *S. aristotelis* before *S. urile* and add *S. gaimardi* before *S. featherstoni* if you want to directly follow the answers. Of course, this addition sequence is arbitrary.

OPTIMAL TREES AND PARSIMONY CRITERIA

An optimal tree is one of many possible trees that meets a particular criterion. Although several criteria exist (i.e., maximum likelihood, parsimony, least square methods for distance data; see Swofford and Olsen [1990] for a readable review), we are concerned only with parsimony as reflected by measurements such as tree length. By this criterion, usually referred to as the maximum parsimony criterion, the optimal tree is the tree that has the shortest length for a particular data set given our assumptions about character evolution.

There may be only a single optimal tree, but there may also be several to many optimal trees. We usually refer to these as the set of equally parsimonious trees

We can consider the distribution of a particular transformation series as "optimal" for a particular tree topology under the maximum parsimony criterion if that distribution provides a narrative explanation of character evolution that minimizes the number of homoplasies and maximizes the number of apomorphies. Homoplasy is not precluded, only minimized for a particular tree topology. Carefully note that we have referred to character optimization given a particular tree topology. Optimization is an *a posteriori* activity; it does not help build trees but is used for evaluating trees that are already built.

The assumptions about character evolution are reflected by several factors, including how we coded the characters (Chapter 3) and how we weight the transformations that have occurred. The weight of a transformation from one character to another represents an assumption on the part of the investigator about the nature of character evolution, which is reflected in the kind of parsimony she picks (actually it's more complicated than this, but this characterization will do for a start). Up to now, we have assumed that all characters have equal weight, which amounts to assigning an equal "cost" of evolution for one step for each transformation within a transformation series. However, you can imagine that the weight of a transformation can be larger or smaller than one.

Swofford and Olsen (1990) review four criteria that provide an introduction to different views on the nature of parsimony.

1. **Wagner parsimony** (Kluge and Farris, 1969; Farris, 1970) treats characters as **ordered** such that the change from one character to another implies change through any intervening characters in the transformation series. Characters are allowed to reverse freely.

2. **Fitch parsimony** (Fitch, 1971) treats characters in a transformation series as **unordered** such that change from one character to another character within a transformation series does not imply changes through possible intervening characters. Characters are allowed to reverse freely.

These two criteria are conceptually simple and do not assume much about the evolutionary process. Their differences can be appreciated by considering the transformation series "0, 1, 2." If a particular tree topology "forces" the hypothesis that "0" evolves to "2," Wagner parsimony would add two steps to the tree, whereas Fitch parsimony would add only one step. Which result is biologically reasonable would depend on the justification for ordering or not ordering the transformation series. An important similarity of the two criteria is that both assign a cost to character reversals. Thus, there is no cost in terms of tree length if the root of the tree is changed. This is quite different from the characteristics of the next two parsimony criteria, where rerooting can have a considerable effect on tree length.

3. **Dollo parsimony** (Farris, 1977) requires every synapomorphy to be **uniquely derived**, i.e., appearing only once on the tree. The synapomorphy may reverse, but once reversed it cannot reappear. Thus, parallel gains of apomorphies are prohibited. Dollo parsimony has been advocated for certain kinds of transformation series (e.g., endonuclease restriction site data; DeBry and Slade, 1985), but a "relaxed" Dollo criterion, which amounts

to assigning a weight to the cost of reversal, might be more appropriate for these data (Swofford and Olsen, 1990). For example, you might assign a weight of "6" to gains of restriction sites, making their parallel evolution very costly in terms of tree length but not impossible. (The weight 6 is not arbitrary but refers to restriction sites six bases in length. The rationale is that the gain of a six-base site is six times less likely than its loss because a single nucleotide change would cause the loss.)

4. **Camin–Sokal parsimony** makes the assumption that character evolution is **irreversible**. This is true, philosophically, because time is irreversible and the reversals are really new apomorphies. But we cannot know this *a priori*. This criterion is rarely used.

Swofford and Olsen (1990) characterize **general parsimony** as an attempt to balance our knowledge of character evolution. You may have knowledge about different sorts of transformation series. This knowledge can be reflected in how you treat the characters. For example, you might have a data matrix composed of both morphological and biochemical characters. You may elect to treat a morphological transformation series of several characters as ordered because you have ontogenetic data that suggest a particular ordering. At the same time, you may have a number of alleles at a particular locus that you treat unordered because there is no clear evolutionary path connecting them that you can determine based on an independent criterion. This transformation series would be treated as unordered. You might have another transformation series comprised of the presence and absence of a particular restriction site and assign a weight to the presence. The result of such an analysis will be a tree supported by characters of various qualities. Tree topologies supported by characters given high weight are harder to reject than tree topologies whose branches are supported by characters weighted equally because of the cost of independently gaining high-weight characters. Thus, differential weighting should be carefully considered before application.

Optimizing Trees

You have already done some optimization in Chapter 2 when you produced trees that had alternate interpretations about the distribution of homoplasies (i.e., Fig. 2.11). Now, we will cover optimization in a more formal manner by considering two basic types, ACCTRAN and DELTRAN (Swofford and Maddison, 1987). ACCTRAN is equivalent to Farris optimization (Farris, 1970) where there is a known ancestor. The etymology of the name is derived from the fact that the procedure ACCelerates the evolutionary TRANsformation of a character, pushing it down the tree as far as possible. The effect is to favor reversals over parallelisms when the choice is equally parsimonious. The tree in Fig. 2.11c is an "ACCTRAN tree." In contrast, DELTRAN DELays the TRANsformation of a character on a tree. The effect is to push the character up the tree as far as possible and to favor parallelisms over reversals when the choice is equally parsimonious. The tree in Fig. 2.11a is a "DELTRAN tree."

ACCTRAN and DELTRAN operate under a general optimality criterion that is concerned with finding the most parsimonious character for each branch of the tree. Such characters are parts of sets of characters termed most parsimonious reconstruction sets (MPR sets), which we will briefly discuss after considering ACCTRAN. The important thing to keep in mind is that both ACCTRAN and DELTRAN will yield the same results when there is no ambiguity, i.e., there are no equally parsimonious choices. Although DELTRAN may favor parallelisms, it is quite possible that if you select DELTRAN every homoplasy in your data set will be a reversal. In the following sections, we concentrate on transformation series that show reversals and/or parallelisms on particular tree topologies, but you should remember that characters showing no homoplasy are optimized in the same manner.

ACCTRAN

Farris (1970) pointed out that although a tree can be built with a Wagner algorithm, this does not always guarantee that the individual characters assigned to the hypothetical ancestors (represented by common nodes) will be optimal given the topology of the tree. Farris then provided an algorithm for optimizing these distributions. His algorithm includes transformation series with two to many characters. We will use a less set-theoretical description employing only binary transformation series. It will serve well enough to give you an idea of how this kind of optimization works. This type of optimization is called a "two pass" method (the Maddison et al. [1984] algorithm discussed in Chapter 3 is an example of a "one pass" method). Using the algorithm, you will assign certain characters to nodes in a pass from the terminal branches to the root (the downward pass) and then reevaluate these assignments in a pass from the root to the terminal branches (the upward pass). Transformation series of three or more states require a bit of knowledge about set theory. Those interested should consult Farris (1970) and Swofford and Madison (1987).

We have broken the process down into three phases, the setup, downward pass, and upward pass. These rules work only in the binary case.

1. The Setup.—On the tree, label each terminal taxon and each ancestor (Fig. 4.5a). Above the taxon name, place its character for a particular character transformation series (Fig. 4.5a).

2. The Downward Pass.—Beginning with the terminal taxa and proceeding toward the root node, assign characters to the ancestral node according to the following rules.

Rule 1.—If both terminal taxa have identical characters, label the node with that character (Fig. 4.5b, R1).

Rule 2.—If they have different characters, label the node with both characters (Fig. 4.5b, R2).

Rule 3.—If one taxon (terminal or ancestral) has a single character (0) and another taxon (terminal or ancestral) has both characters (0,1), label their common node with the **majority** character (Fig. 4.5b, R3). If both characters are equally common (possible with polytomies), then assign both characters.

Rule 4.—If both taxa have both characters (i.e., both are 0,1), then their common node will have both characters (see Fig. 4.6a).

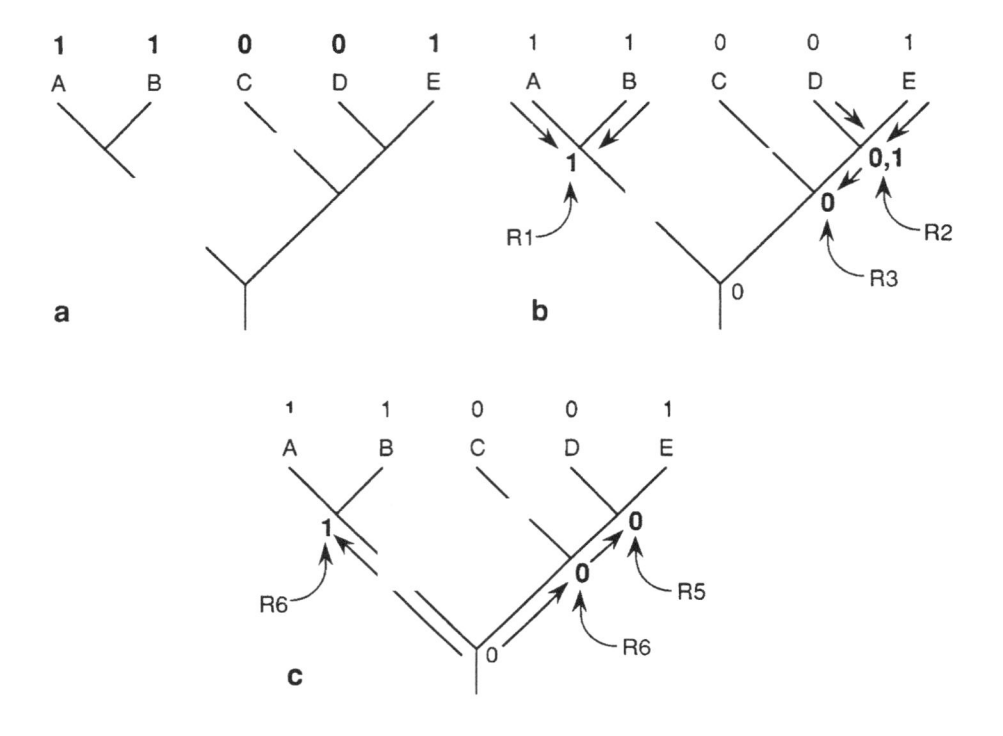

Fig. 4.5.—ACCTRAN optimization. a. The setup. b. The downward pass. c. The upward pass. R1–R6 are the application of the rules discussed in the text.

3. The Root Node.—When you get to the root, label it with the ancestral character (whether it is 0 or 1 or 0,1) regardless of what the last two taxa may indicate. In the "real world," this assignment would have been determined before you began by using the Madison et al. (1984) algorithm discussed in Chapter 3. We have arbitrarily selected "0" in these examples.

4. The Upward Pass.—Begin working upward from the root node to the terminal taxa. The following rules apply to the upward pass.

Rule 5.—If a character assignment at a node is 0,1, then assign the character of the node immediately below it (Fig. 4.5c, R5). (Equivalent to Rule 3, but going the other way.)

Rule 6.—If a character assignment is either 1 or 0, then do not change that character assignment, even if the node below it is different (Fig. 4.5c, R6).

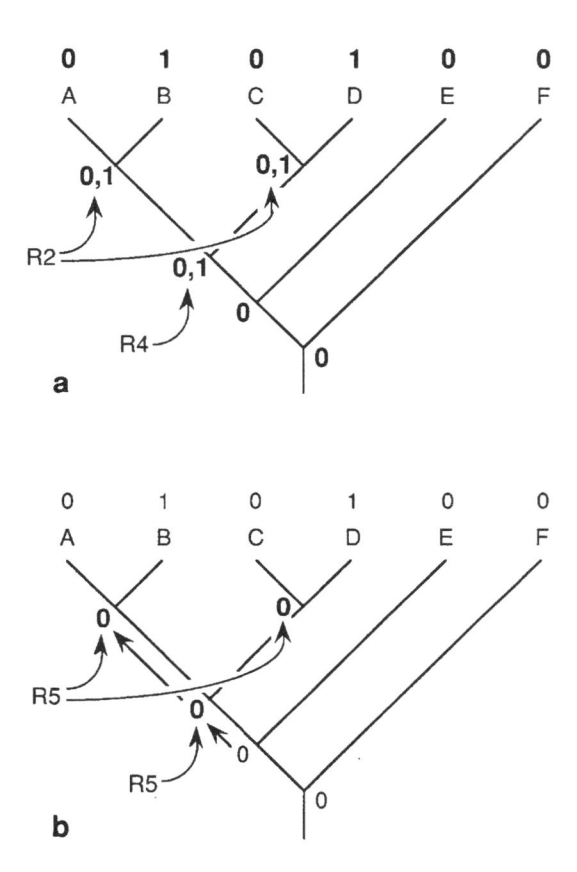

```
    0   1   0   1   0   0
    A   B   C   D   E   F
     0,1    0,1
  R2      0,1
            0
          R4
  a         0


    0   1   0   1   0   0
    A   B   C   D   E   F
      0      0
  R5       0
           0
         R5
  b         0
```

Fig. 4.6.—State assignments on an ACCTRAN optimization pass. **a**. A downward pass. **b**. An upward pass.

Example 4.3.—Simple character assignments.

In Fig. 4.6a, we show the character assignments on the downward pass and in Fig. 4.6b the character assignments on the upward pass. Each assignment is associated with the rule applied.

Example 4.4.—Complex character assignments (Fig. 4.7).

As in Example 4.3, we illustrate character assignments for the downward pass (Fig. 4.7a) and the upward pass (Fig. 4.7b).

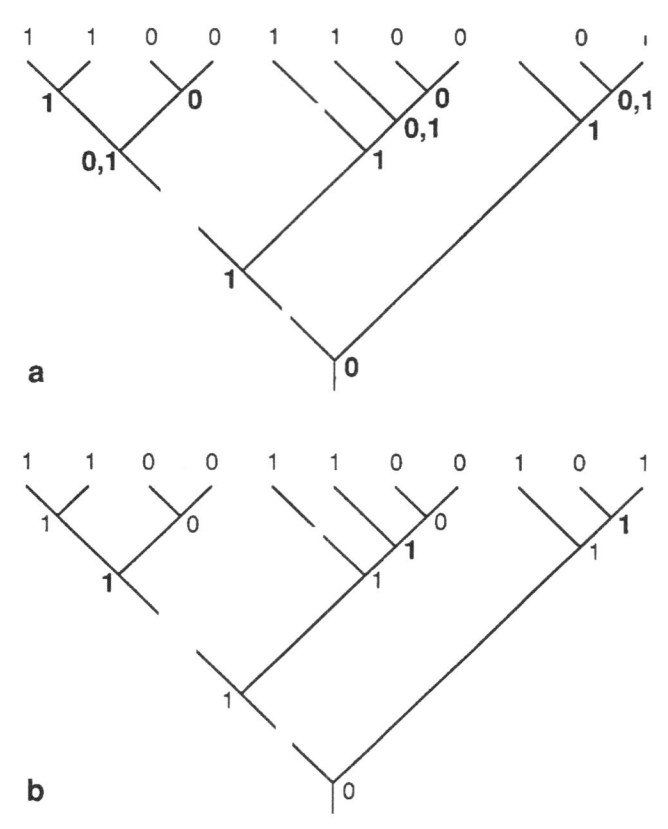

Fig. 4.7.—Another example of ACCTRAN optimization. a. State assignments after the downward pass. b. Nodes optimized after the upward pass.

ACCTRAN Exercises

Use the information in the accompanying figures to optimize the character distributions.

EXERCISE 4.7.—Use Fig. 4.8 for optimization and label the root node "0."

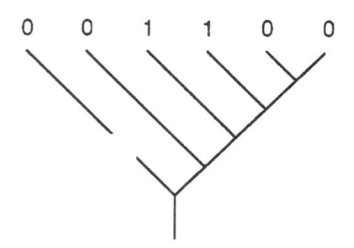

Fig. 4.8.—Tree for ACCTRAN optimization (Exercise 4.7).

TREE BUILDING AND OPTIMIZATION

EXERCISE 4.8.—Use Fig. 4.9 for optimization and label the root node "0."

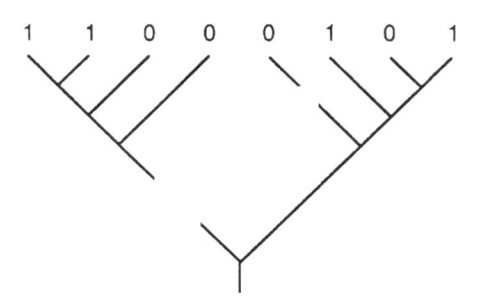

```
1   1   0   0   0   1   0   1
```

Fig. 4.9.—Tree for ACCTRAN optimization (Exercise 4.8).

EXERCISE 4.9.—Use Fig. 4.10 for optimization and label the root node "0."

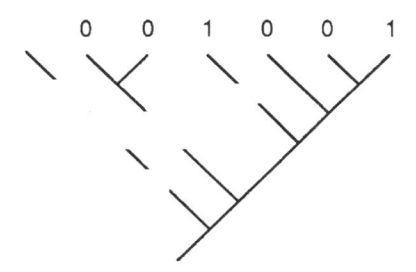

```
    0   0   1   0   0   1
```

Fig. 4.10.—Tree for ACCTRAN optimization (Exercise 4.9).

Discussion

ACCTRAN is a special case of a more general way to optimize characters on a tree (Swofford and Maddison, 1987). It will always insure that the length of the tree is minimized given a particular character matrix, but it will usually only show one interpretation of how the characters evolved (a formal proof of Farris's algorithm is provided by Swofford and Maddison [1987]). That is, it favors reversals over repeated origins when the choice is equally parsimonious. We can see this by examining the very simple tree in Fig. 4.11a.

When you optimize the characters on the tree using ACCTRAN, the result shows a reversal from 1 to 0 (Fig. 4.11b) (if this is not obvious, apply the ACCTRAN rules). The tree has a length of two steps. However, there is another equally parsimonious tree that interprets the evolution of a character coded "1" as a homoplasy (Fig. 4.11c). The problem is ACCTRAN will never allow you to find this tree.

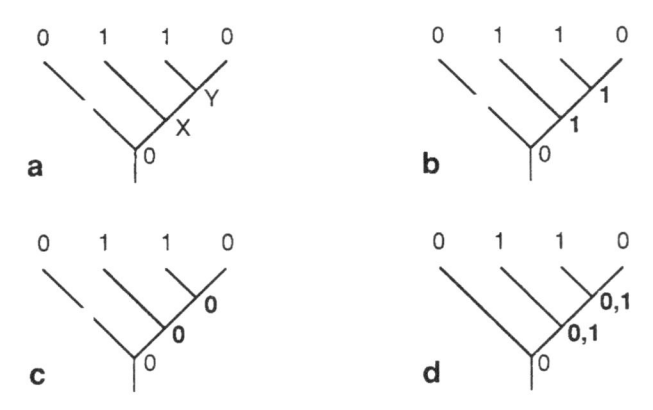

Fig. 4.11.—Four views of a tree. **a**. The tree with characters for the terminal taxa and the root node. **b**. The tree optimized by ACCTRAN. **c**. Another most parsimonious interpretation of the characters X and Y. **d**. The most parsimonious resolution (MPR) set, combining the information in b and c.

Let us think about the character assignments of these two trees. Obviously, they differ. If we combine them into one tree, ancestor Y_i has a state set 0,1 rather than just 1 or 0 (Fig. 4.11d). Because this character set is the set that combines both possibilities, it is termed the "MPR set" for that particular ancestor (Swofford and Maddison, 1987). Because the MPR set contains two characters, we can assume that there are two equally parsimonious interpretations of character evolution. In Fig. 4.11b, the evolution of character 1 is accelerated; in Fig. 4.11c, the evolution of 1 is delayed.

Finding MPR Sets

We find MPR sets by evaluating the possible character assignments for each interior node (i e., each ancestor), which is accomplished by rerooting the tree at each of these nodes, beginning with the node closest to the root and working toward the terminal branches.

1. Proceeding up the tree (Fig. 4.12a) from the root (X), reroot the tree at the first ancestral node (Y_i) (Fig. 4.12b).

2. Optimize the character assignment of this node by using the downward pass as specified in ACCTRAN. The MPR set will be composed of the majority character using Rule 3. Note that neither character is in the majority, so both (i.e., 0,1) are assigned to Y_i (Fig. 4.12b).

3. Proceed to the next ancestral node (Z) and reroot the tree (Fig. 4.12c). Follow step 2. To reduce redundant calculations, use the MPR set of any ancestor whose MPR set has already been determined. In Fig. 4.12c, note that we have used the MPR set of Y_i as well as the character sets of C and D to determine the MPR set of Y_i. This is a short cut. You are finished after all ancestral nodes have been evaluated.

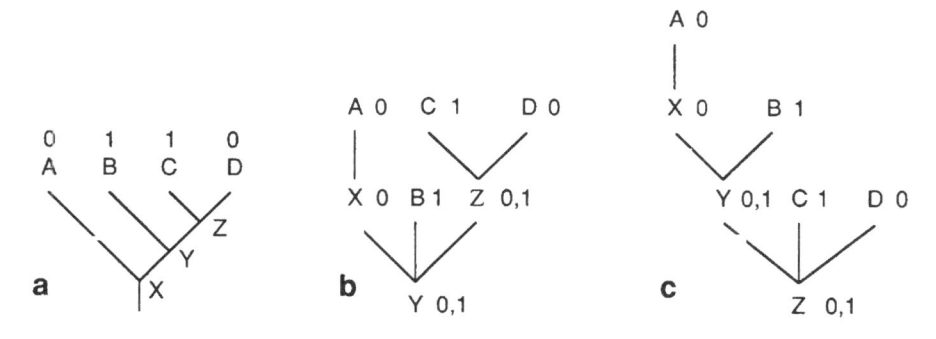

Fig. 4.12.—Finding the MPR set for clade ABCD. **a.** Tree. **b.** Tree rerooted at Y and optimized using ACCTRAN. **c.** Tree rerooted at Z and optimized using ACCTRAN.

DELTRAN

DELTRAN is the opposite of ACCTRAN. It favors parallelisms when given the chance to do so. To implement the DELTRAN option, you must be able to find the MPR sets for each ancestral node.

1. Determine the MPR sets for each ancestral node by using the procedure outlined above, and place them on the tree.
2. Implement DELTRAN as in the upward ACCTRAN pass, using Rules 5 and 6.

Example 4.5.—MPR sets and DELTRAN optimization.

We will use the tree in Fig. 4.13a to 1) determine the MPR sets for each ancestor and 2) optimize the tree using DELTRAN. Note that the number of terminal taxa used decreases as the MPR sets for their common nodes are included as we go up the tree. For example, A and OG are represented in Fig. 4.13c by S and its MPR set (0).

1. We determine the MPR sets for each hypothetical ancestor, working from the root up the tree (Fig. 4.13b–g).
2. We place the MPR sets on the original tree (Fig. 4.14a).
3. Finally, we optimize the tree in an upward pass using Rules 5 and 6 (Fig. 4.14b).

DELTRAN Exercises

EXERCISE 4.10.—Using the matrix in Table 2.2 (p. 19) and the tree in Fig 2.11a (p. 21), optimize transformation series 3 and 4 using DELTRAN optimization. Then optimize the tree using ACCTRAN. Compare your results with the trees in Fig. 2.11a, c. Use only the topology; disregard the characters on the tree.

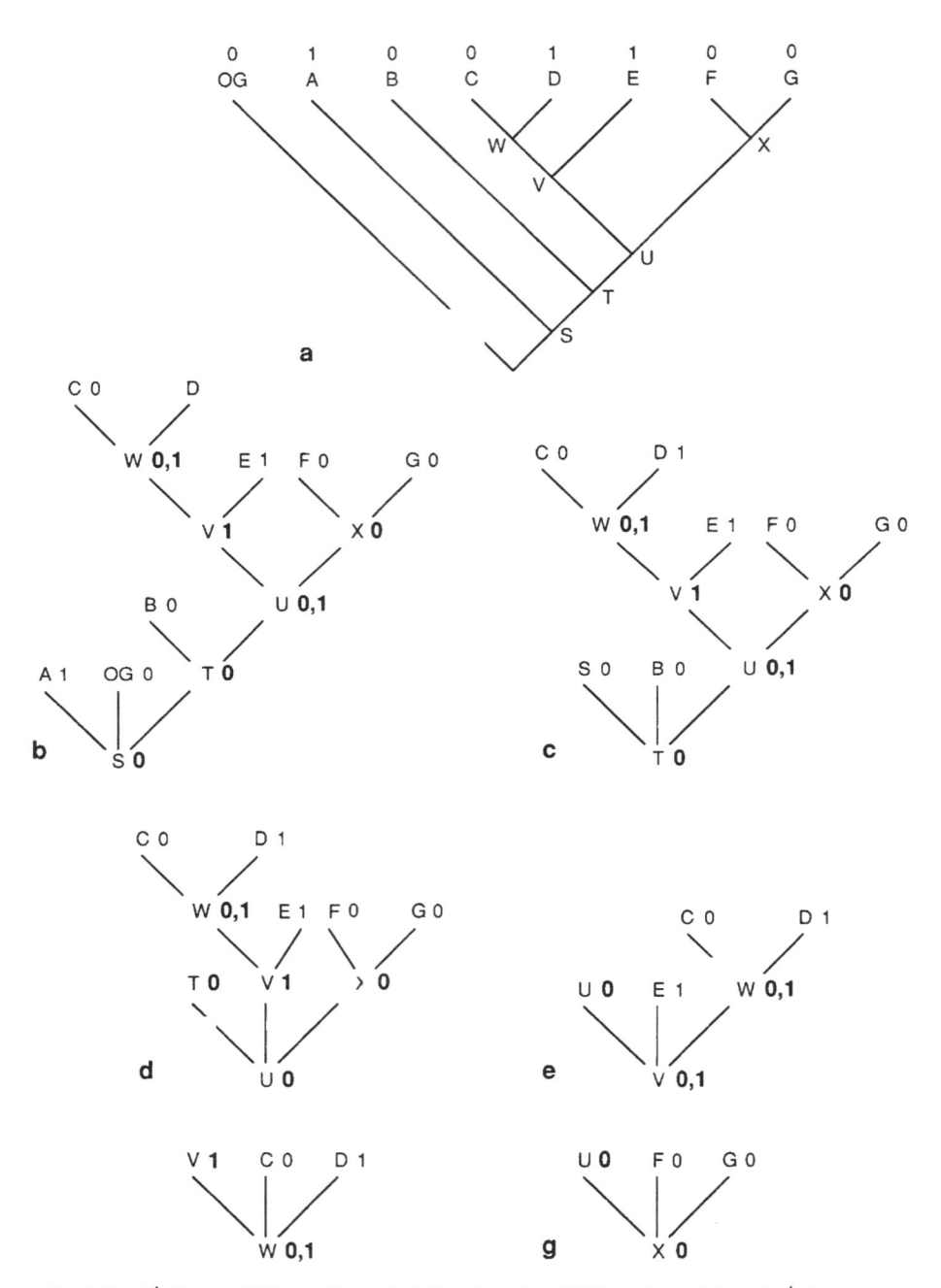

Fig. 4.13.—Finding the MPR sets (Example 4 5). a. Tree. b–g. MPR sets for each hypothetical ancestor.

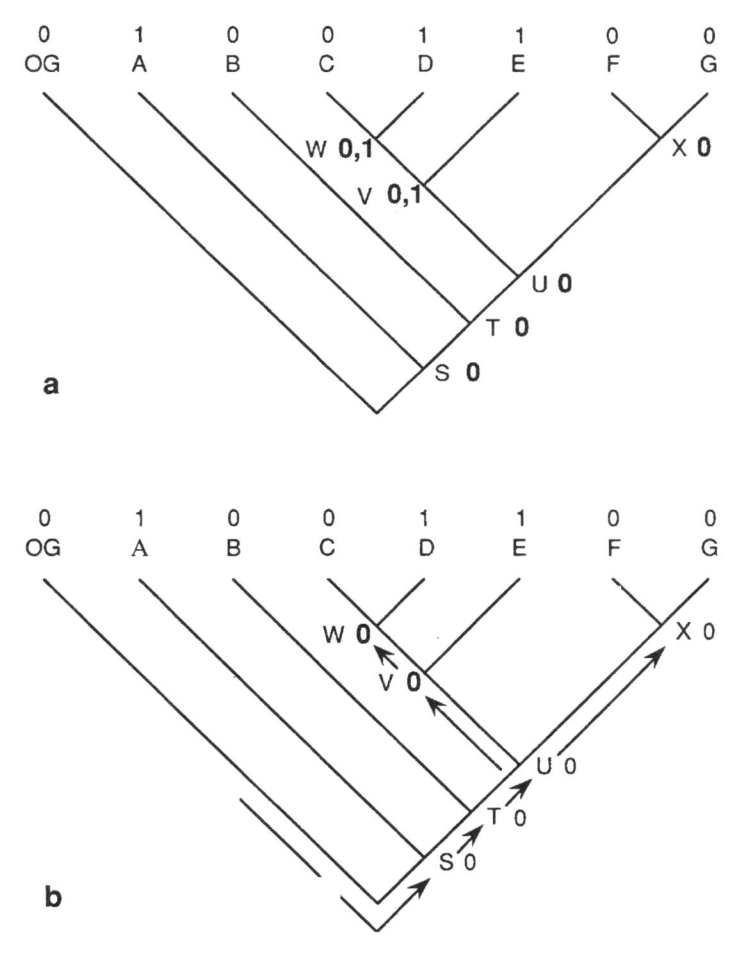

Fig. 4.14.—DELTRAN optimization using the MPR sets in Fig 4 13 (Example 4.5). **a.** The original tree with MPR sets placed at the nodes. **b.** Each node optimized using DELTRAN.

EXERCISE 4.11.—Use the tree in Fig. 4.15 to find the MPR set for each ancestor. Use these sets to optimize the tree with DELTRAN. Then optimize using ACCTRAN.

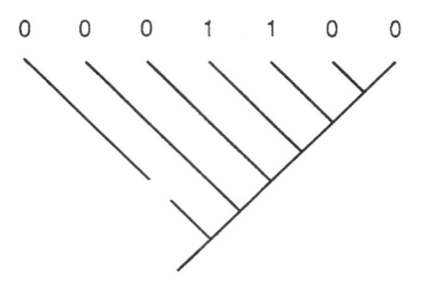

```
0   0   0   1   1   0   0
 \   \   \   \  /   /   /
  \   \   \   \/   /   /
   \   \   \  /\  /   /
    \   \   \/  \/   /
     \   \  /\  /   /
      \   \/  \/   /
       \  /\  /   /
        \/  \/   /
        /\  /   /
```

Fig. 4.15.—Tree for MPR sets and DELTRAN optimization (Exercise 4.11).

Current Technology

Modern computer packages like PAUP (Swofford, 1990) are composed of programs containing a wealth of options. Actual analyses may take seconds, hours, or days. Interestingly, most of the time spent in analyzing the data is not spent on tree building *per se* but on evaluating tree topologies using a relevant optimality criterion. The approach to finding the optimal tree can differ considerably from traditional methods. Swofford and Olsen (1990) characterize three general approaches: 1) exhaustive searches, 2) branch-and-bound searches, and 3) heuristic searches. The first two can provide an exact solution if the data matrix is small enough. By exact solution we mean that the resulting tree or set of trees will be the shortest tree(s) for the data given maximum parsimony as the criterion. The heuristic search will find a short tree(s), but there is no guarantee that the tree will be the shortest. Heuristic searches are performed on much larger data sets instead of using searches guaranteed to find an exact solution because for many large data sets there are no other alternatives (see below). The following abbreviated characterization of these three approaches is abstracted from Swofford and Olsen (1990).

1. **Exhaustive search.**—An exhaustive search consists of evaluating the data over all possible trees. Because there are no other possible topologies unevaluated, the shortest tree or set of trees will be found. As you will appreciate when you get to Chapter 6, the number of possible tree topologies increases at a very rapid rate. An exhaustive search on 10 taxa is very efficient, but the same procedure performed on 25 taxa may be prohibitive in terms of computer time.

2. **Branch-and-bound search.**—Branch-and-bound algorithms can provide an exact solution for a larger number of taxa than exhaustive search can because the search procedure it employs has a provision for discarding trees without evaluating them if they meet certain criteria. Hendy and Penny (1982) first introduced branch-and-bound algorithms to phylogenetic analysis, and a good description of what they do can be found in Swofford and Olsen

(1990). In essence, branch-and-bound works with a tree of possible trees termed a search tree (Fig. 4.16). The root of the search tree is a three-taxon tree (unrooted so it is the only possible tree). From this root tree, other possible trees are derived by adding additional taxa at each place where they might be added. So, D could be added in one of three places to yield three trees, and E could be subsequently added in any one of five places to these three trees, yielding the 15 trees in Fig. 4.16. Now, let us place an upper bound on the search. The upper bound is the maximum length a tree can obtain before it is eliminated. As we go from the root of the search tree towards its tips, we will only reach the tips along a certain path if our upper bound is not exceeded. If we exceed the upper bound, we cut off that branch and do not evaluate any additional branches connected to it. Rather, we backtrack down the search tree and proceed up another branch. If we find a tree that is better than the upper bound (i.e., a tree shorter than the one we used for the initial upper bound), we adopt this superior

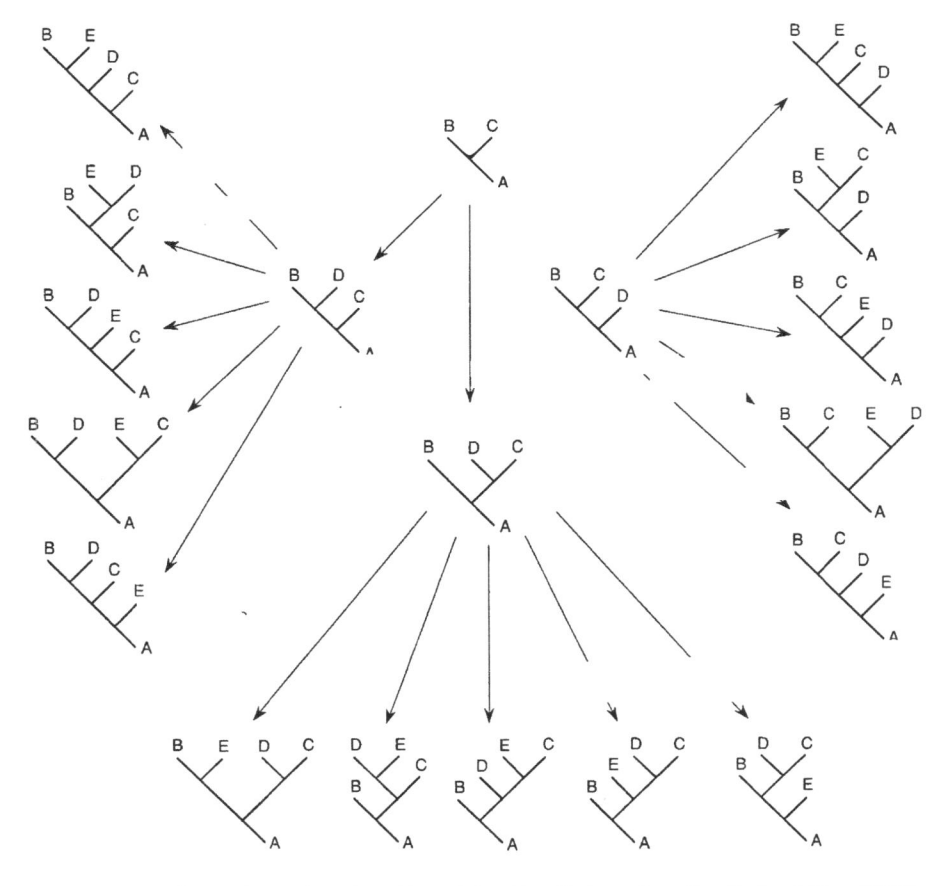

Fig. 4.16.—Trees illustrating the branch-and-bound algorithm. Redrawn from Swofford and Olsen (1990).

standard for all subsequent analyses. Thus, we might eliminate additional trees that would be acceptable given the old, more relaxed, upper bound. Trees that are not eliminated are candidates for status as most parsimonious trees. If a better upper bound is found, these trees are subject to further evaluation. What is left is the set of optimal trees. This is the trick: by not evaluating all the branches to all the tips, we avoid an exhaustive search without suffering loss of precision because we do not have to evaluate tens, hundreds, thousands, or even millions of possible tree topologies that are not close to the optimal solution. Obviously, we need to pick an initial upper bound before the analysis begins. This can be done by picking a random tree and calculating its tree length. For our example of five taxa (Fig. 4.16), we would pick any of the 15 trees, calculate its tree length, and begin. If that tree turned out to be the worst possible tree, then we would expect to quickly find a new, and lower, upper bound. If the tree turned out to be the most parsimonious tree (a 15-to-1 shot), then the upper bound would be so low as to quickly eliminate many of the less parsimonious trees. PAUP uses the next category of methods, heuristic methods, to pick a near optimal tree whose length is used as the initial upper bound. These and other tricks speed up the process (Swofford, 1990; Swofford and Olsen, 1990).

3. **Heuristic search.**—Heuristic search begins by building a tree. One common way to do this is via the Wagner algorithm. A number of other approaches can also be used, depending on the computer package. The tree obtained may be optimal, especially if the data are relatively free of homoplasy. However, it may not be optimal. The manner in which taxa are added to the growing tree constrains the topologies that can be built from taxa added subsequently, ending in a "local optimum," i.e., a tree that is the "optimum" given the constraints provided by the way the taxa were added but not necessarily the best tree if all possible trees were evaluated. This situation is rather analogous to proceeding along a reasonable path in a search tree but not having the option of backtracking to try other possibilities. Heuristic search routines attempt to circumvent local optima by branch swapping. Branch swapping involves moving branches to new parts of the tree, producing new tree topologies. The data are optimized on the new topology, and if the tree is shorter, it is subjected to additional branch swapping. Eventually, these rounds of branch swapping will lead to an optimal tree or set of optimal trees. Several pitfalls can be encountered, and strategies for avoiding them must be adopted (Swofford and Olsen, 1990).

CHAPTER NOTES AND REFERENCES

1. Hennig argumentation is covered in some detail by Hennig (1966), Eldredge and Cracraft (1980), and Wiley (1981a). It works well with relatively small data sets with little homoplasy. However, we continue to be amazed to see how few homoplasies it takes in a data matrix to yield more than one most parsimonious tree.

2. Those already familiar with data analysis will note that we have not discussed alternative approaches to phylogeny reconstruction such as compatibility analysis, analysis of distance data (Farris, 1972, 1981, 1985, 1986; Felsenstein, 1984, 1986; Hillis, 1984,

1985), likelihood methods (Felsenstein 1973a,b, 1981; Thompson, 1986), and bootstrap methods (Felsenstein, 1985; Sanderson, 1989). You should consult the literature about these controversies.

3. Those interested in analysis of molecular data should consult Hillis and Moritz (1990).

4. Finding the MPR set for transformation series with more than two characters is not so easy and requires more set manipulations. Swofford and Maddison (1987) outline the procedure and provide proofs of both Farris's (1970) optimization algorithm and their own algorithms. However, the paper is very technical. Farris (1970) gives a clear example of how to optimize a multicharacter transformation series using ACCTRAN.

5. Remember that ACCTRAN and DELTRAN will give the same solutions for character evolution only if it is equally parsimonious to do so. The majority of transformation series in a particular data matrix may not contain MPR sets with two or more characters, and both ACCTRAN and DELTRAN will give the same values. Also, consider the situation where we may have many equally parsimonious decisions to make, each involving a different transformation series. We might prefer one transformation series to be interpreted as being subject to parallel evolution (DELTRAN) while another is considered as being subject to reversals (ACCTRAN). In any one analysis, however, we cannot "mix and match." Given a particular topology, we can only opt for a uniform DELTRAN or a uniform ACCTRAN optimization. The extent to which a particular investigator opts for the interpretation that a particular transformation series is "more likely" to be subject to parallelism than to reversal does not depend on an optimization routine but on specific assumptions about the evolutionary constraints placed on the descent of particular kinds of characters.

KU MUSEUM OF NATURAL HISTORY SPECIAL PUBLICATION No. 19

CHAPTER 5

TREE COMPARISONS

In the early analyses of groups using phylogenetic techniques, only rarely did investigators report more than one tree, frequently because the data were "clean," i.e., relatively free from homoplasy, but sometimes because the investigator did not take the time to explore the data set for additional trees. Computers have changed this. A computer will take the time to find as many alternative trees as the program it employs can identify. As a result, more trees are being reported, and measures are needed to sort them out. In this chapter, we will introduce you to 1) some basic tree measures and 2) consensus techniques, i.e., techniques for exploring the similarities and differences between trees with differing branching structures.

SUMMARY TREE MEASURES

Summary tree measures are designed to give you a set of basic "statistics" you can use to evaluate the differences between two trees that have been generated from the same data set. They are relative measures that say something about the basic structure of the data, the optimization technique employed, and the algorithm used. They are not very useful for comparing the results of two different data sets for the same taxa.

Tree Length

We have been working with tree length since Chapter 2. Tree length is calculated by summing the number of character changes along each branch and internode of the tree. For a given data set, the "best tree" is defined as the tree of shortest length because, as we discussed in Chapter 2, it provides the most parsimonious description of all homology and polarity arguments taken together. We hope that the shortest tree will also be the best estimate of the actual common ancestry relationships of the taxa analyzed. Two or more shortest trees represent equally parsimonious solutions to the same data set for a particular analysis. There are two kinds of equally parsimonious trees: 1) the set of trees that show the same common ancestry relationships but differ in character interpretation and 2) the set of trees that differ in topology and thus represent different views of the common ancestry relationships. These two types of equally parsimonious trees have different qualities. For systematics and biogeography, equally parsimonious trees that differ in character interpretations but have identical topologies do not affect any subsequent analysis that takes advantage of the tree topology. However, the differences among these trees might be important when dealing with tests of evolutionary mechanisms and character evolution. Trees of differing topology directly affect subsequent comparative analysis and can lead to problems in presenting taxonomies because there are at least two different views of the common ancestry relationships among the taxa involved in the analysis.

Consistency Indices

Kluge and Farris (1969) introduced consistency indices as measures of how transformation series and entire data matrices "fit" particular tree topologies. Nontechnically, transformation series with little or no homoplasy have high consistency index values (1.0 is the highest value possible), whereas those that show considerable homoplasy have low values. Aside from tree length, consistency indices are the most commonly reported values for trees. Since their introduction in 1969, several modified consistency indices have been suggested. We will cover the basics of these measures using the data matrix in Table 5.1 and the trees in Fig. 5.1. We will first cover the measures for individual transformation series and then consider the ensemble measures for entire data matrices.

Table 5.1.—Data matrix for the hypothetical clade A–E and its sister group OG.

Taxon	\multicolumn{8}{c}{Transformation series}
	1	2	3	4	5	6	7	8
OG	0	0	0	0	0	0	0	0
A	1	0	0	0	0	1	0	1
B	1	1	1	0	1	0	1	0
C	1	0	1	1	1	0	0	0
D	1	1	1	0	1	1	1	0
E	1	1	1	1	1	1	1	1

The consistency index of a transformation series of discrete characters (c or ci) is the ratio of the minimum amount of changes (steps) it might show (m) and the amount of change (steps) it does show on a particular tree (s):

$$c = \frac{m}{s}$$

For binary transformation series, $m = 1$; for multicharacter series, the m-value will equal the minimum number of steps possible. For example, a transformation series with three characters (0, 1, and 2) would evolve a minimum of two steps. For simplicity, we use only binary transformation series in both the examples and the exercises.

Consider TS 1 of Table 5.1. As a binary transformation series, $m = 1$. Now look at Fig. 5.1. There has been one transformation from 1-0 to 1-1 and no reversals from 1-1 to 1-0. The number of steps (or length) of this series on the tree is $s = 1$. The consistency index for this transformation series is

$$c = \frac{m}{s} = \frac{1}{1} = 1.0.$$

Now consider TS 8. As a binary transformation series, $m = 1$. On the tree, however, character 8-1 has arisen twice, so $s = 2$ and

$$c = \frac{m}{s} = \frac{1}{2} = 0.50.$$

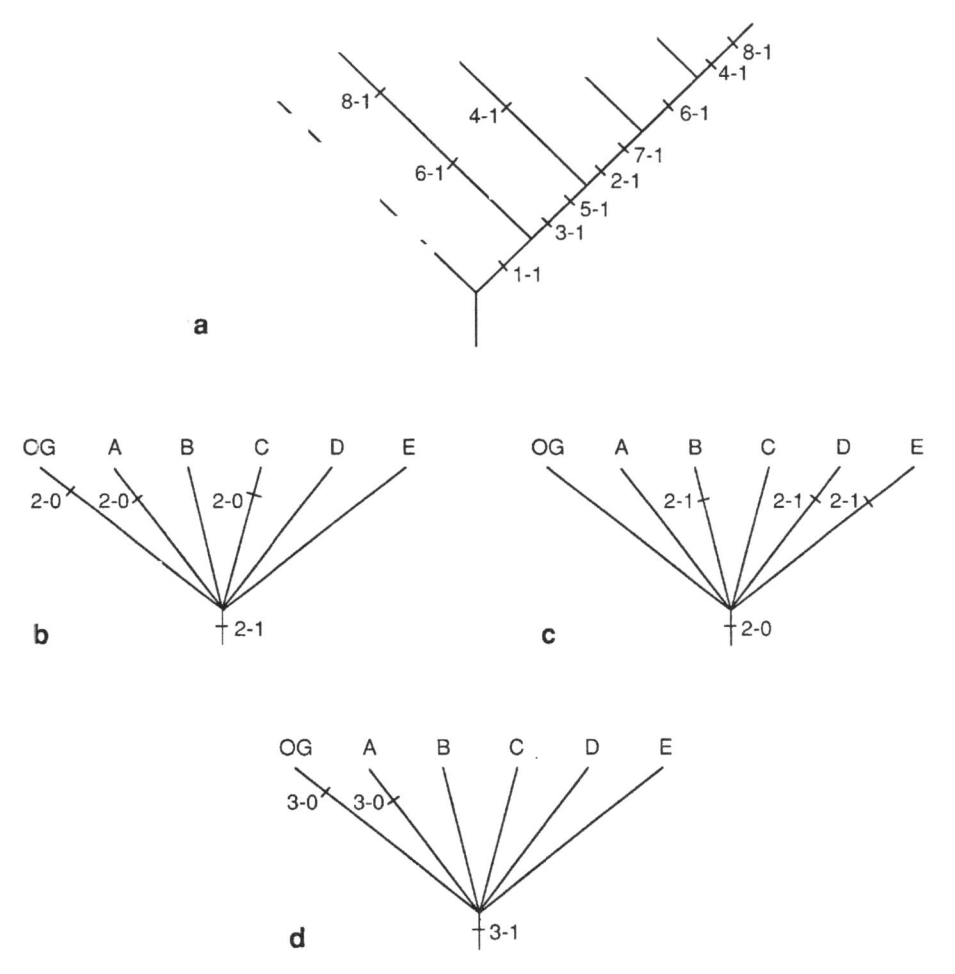

Fig. 5.1.—a. Tree for the hypothetical clade A–E and its sister group OG. b, c. Polytomies for TS 2. d. Polytomy for TS 3.

Now consider TS 6, another binary character with s = 1. Figure 5.1 shows that 6-0 has given rise to 6-1 twice, once in taxon A and once in the ancestor of taxa D and E. The consistency index for this character is

$$c = \frac{m}{s} = \frac{1}{2} = 0.50.$$

Although the consistency indices for TS 6 and TS 8 are identical, there is a difference in their performance on the tree. TS 8 shows total homoplasy in the evolution of character 8-1, whereas TS 6 shows only partial homoplasy in the evolution of 6-1. That is, a derived character in TS 6 does lend support to the tree topology in Fig. 5.1, whereas derived characters in TS 8 lend no support to the tree topology at all. Yet, they have exactly the same consistency index values.

To overcome this problem, Farris (1989b) introduced the rescaled consistency index (actually first introduced in Farris' Hennig 86 [1989a]). The rescaled consistency index (rc) is the product of the original consistency index (c) of Kluge and Farris (1969) and the retention index (r) of Farris (1989b). The retention index measures the fraction of apparent synapomorphy to actual synapomorphy. To calculate the retention index we need the s-value, the m-value, and a new value called the g-value, for each transformation series. The g-value is a measure of how well a transformation series might perform on any tree, i.e., how many steps would it take to explain evolution within the transformation series under the worst possible condition. The case of any tree is the case of the phylogenetic "bush," a polytomy involving all taxa (including the outgroup).

Consider TS 2 in Table 5.1. In the worst possible circumstances (the phylogenetic bush), what is the minimum number of steps needed to explain the evolution of the characters in this transformation series? In Fig. 5.1b we have a bush (including the outgroup). If we assign 2-1 to the root, then 2-0 would evolve three times, creating three steps. If we assign 2-0 to the root, then 2-1 requires three steps (Fig. 5.1c). So, assigning 2-0 or 2-1 to the root results in the same number of steps, and g = 3. Any fewer number of steps would require us to group taxa, and this would not represent the minimum number of steps in the worst case (the g-value) but would instead represent the minimum number of steps in the best case (the m-value).

Now consider TS 3. If we assign 3-1 to the root (Fig. 5.1a), then 3-0 has to evolve twice (g = 2). If we assign 3-0 to the root, then 3-1 has to evolve four times. Parsimony prefers two steps to four steps, so the best we can do under the worst tree topology is two steps.

For binary characters we can apply a short-cut to determine the g-value. For a transformation series with binary characters, the g-value is the smallest of the values for the occurrence of the two characters. TS 3 contains two 3-0 values and four 3-1 values, and g = 2. The number of 0-values and 1-values in TS 2 is equal, so the g-value equals this value (g = 3). There is only a single 1-0 in TS 1, so g = 1.

The retention index, r, is defined by Farris (1989b) as

$$r = g - \frac{s}{g} = m. \quad \frac{g-s}{g} \quad [SEE\ ERRATA.]$$

When we calculate some of the r-values for the transformation series in Table 5.1 using the tree in Fig. 5.1a, we find the following values:

$$TS\ 2:\quad r = \frac{3-1}{3-2} = \frac{2}{2} = 1.00$$

$$TS\ 6:\quad r = \frac{-2}{-1} = \frac{\frac{1}{2}}{} = 0.50$$

$$TS\ 8:\quad r = \frac{-2}{-1} = \frac{0}{1} = 0.00.$$

Note the contrast between TS 6 and TS 8. Although there is no actual synapomorphy in TS 8, there is *apparent* synapomorphy (i.e., reflected by the original coding of the homoplasies represented by the code "8-1"), and the retention index for this transformation series is zero. Some synapomorphy is present in TS 6, so its retention index is greater than zero.

The rescaled consistency index (rc) for a transformation series is the product of the consistency index and the retention index:

$$TS\ 2:\ rc = (1)(1) = 1.00$$
$$TS\ 6:\ rc = (0.5)(0.5) = 0.25$$
$$TS\ 8:\ rc = (0)(0.5) = 0.0.$$

Note the relative performance of the two transformation series showing homoplasy.

Finally, consider TS 1. This transformation series presents the special case $m = s = g = 1$. Obviously character 1-1 represents a synapomorphy for the group *given the assumption that 1-0 is plesiomorphic*. But what if 1-1 were plesiomorphic and 1-0 were autapomorphic for the outgroup? There is no way to specify this, so we will calculate $r = 0$ and $rc = 0$. Biologically, we "know" that 1-1 is a synapomorphy, and thus $r = 1$ and $rc = 1.0$. But algorithmically, the only way to produce this result would be to include another outgroup. Biologically, we can interpret this transformation series directly. A character shared by all members of the ingroup is consistent with all possible topologies and thus is not informative. Likewise, autapomorphies will yield $r = 0$ and $rc = 0$ because they are also consistent with all possible topologies (see Exercise 5.1).

Ensemble Consistency Indices

Ensemble consistency indices can be used to examine the relationship between an entire data matrix and a particular tree topology. The most commonly used index is the ensemble or "overall" consistency index (CI) of Kluge and Farris (1969), which can be calculated very simply for a binary matrix by taking the ratio of the number of data columns and the length of the tree. In general, a high CI indicates that the data matrix "fits" the tree well (i.e., contains little homoplasy for the particular tree topology), whereas a low CI does not.

Although the CI is the most commonly reported measure of fit between a character matrix and a tree (Sanderson and Donoghue, 1989), it suffers from some problems. Brooks et al. (1986) showed that the CI was influenced by the number of autapomorphies that are present in the data matrix. The actual support for a particular tree may be less than the apparent support because autapomorphies ($c = 1.0$) contribute to the CI without supporting any particular tree topology. One solution to this problem would be to eliminate autapomorphies from the calculations (Carpenter, 1988). Another problem is the negative relationship between the CI and the size of the data set (Archie, 1989), rendering the CI suspect when comparing different groups of taxa or the same taxa for different data matrices of various sizes.

In an effort to address some of these problems, Farris (1989b) suggested that ensemble consistency indices be calculated using rescaled values. These calculations are demonstrated in Example 5.1.

Example 5.1.—Rescaled consistency indices.

We will use Table 5.1 and Fig. 5.1a as the basis for calculating the ensemble values of the retention index (R) and the rescaled consistency index (RC). To do this, we obtain sums of the s-, m-, and g-values for each individual transformation series. These values are shown in Table 5.2. Note that because we are working with binary transformation series, m = 1 for each data column and thus M = 8 (the number of data columns). If we sum each s-value, we arrive at S = 11, the length of the tree. The sum of the minimum number of steps for each transformation series given a polytomy (the g-value) yields G = 18. We can now calculate the CI, R, and RC:

$$CI = \frac{M}{S} = \frac{8}{11} = 0.727$$

$$R = \frac{G-S}{G-M} = \frac{18-11}{18-8} = 0.700$$

$$RC = (0.727)(0.700) = 0.509.$$

Table 5.2.—Some values* used to calculate rescaled consistency indices (Example 5.1).

TS	m		g	ci		rc
1	1		1	1.00	0/0	0/0
2	1	1	3	1.00	1.00	1.00
3	1	1	2	1.00	1.00	1.00
4		2	2	0.50	0.00	0.00
5		1	2	1.00	1.00	1.00
6	1	2	3	0.50	0.50	0.25
7	1	1	3	1.00	1.00	1.00
8	1	2	2	0.50	0.00	0.00
Totals	8	11	18			

* m = no. changes a character might show on a tree; s = no. changes a character does show on a tree; g = minimum no. steps for each TS given a polytomy; ci = character consistency index; r = character retention index; rc = character rescaled consistency index.

The RC excludes characters that do not contribute to the "fit" of the tree, preventing inflation of the CI. The RC is superior to the method of excluding autapomorphies because it not only excludes these characters, but also excludes totally homoplastic characters, preventing them from artificially inflating the measure of fit (e.g., TS 8), while allowing characters that are partly homoplastic but partly supportive of the tree topology (e.g., TS 6) to contribute to the ensemble value.

The F-Ratio

The F-ratio (apparently first presented in the LFIT function of PHYSYS, a computer program developed in 1982 by M. F. Mickevich and J. S. Farris) is another commonly reported tree statistic. It is a measure of the differences between the phenetic differences between taxa and the path-length distances between these taxa as shown by a particular tree. Low F-ratios are considered superior to high F-ratios. To calculate the F-ratio we need 1) a data matrix, 2) a matrix of phenetic differences, and 3) a matrix of path length (patristic) distances. The F-ratio is then calculated in the following manner.

Example 5.2.—F-ratio calculations.

1. From the data matrix (Fig. 5.2a), solve the phylogenetic relationships of the group (Fig. 5.2b).

2. From the data matrix, prepare a matrix of phenetic distances (Fig. 5.2c). We do this by constructing a taxon × taxon matrix. The phenetic difference between two taxa is the sum of their absolute differences. So, the distance from OG to A is the difference between the OG character vector and the A character vector, which is 2. The distance between B and D is 3, etc.

3. From the phylogenetic tree, prepare a matrix of path-length (patristic) distances (Fig. 5.2d) by preparing another taxon × taxon matrix. This time we count the steps along the path specified by the tree. For example, there are two steps from OG to A and from A to B. The difference between the phenetic and path-length distances becomes apparent when we compare the A to D path. Phenetically, the difference between A and D is 3. Patristically, the path-length difference is 5 because the tree interprets character 5-1 as a homoplasy.

4. Calculate the f-statistic. The f-statistic is the sum of the difference matrix. The difference matrix is a matrix of differences between the phenetic and path-length matrices (Fig. 5.2e). For example, there is no difference in the A/B cell, so the value is "0," whereas there is a difference of 2 in the A/D cell, so the value is "2."

5. The F-ratio is calculated as the ratio of the f-statistic and the sum of the phenetic distance matrix. To sum the phenetic distance matrix, you simply sum each column and then add the column sums together. In this case, the F-ratio is

$$\text{F-ratio} = \left(\frac{f\text{-statistic}}{\text{patristie matrix}}\right)(100)$$

$$= \left(\frac{2}{28}\right)(100)$$

$$= 7.14.$$

The F-ratio has two major drawbacks. First it is not entirely clear what the F-ratio is supposed to measure because we do not really know what a phenetic difference means relative to a phylogenetic tree. Second, although the F-ratio can distinguish autapomorphies from internal synapomorphies (i.e., synapomorphies not associated with the root internode), it treats synapomorphies at the root internode as if they were autapomorphies because the calculation of the F-ratio is independent of the placement of the root.

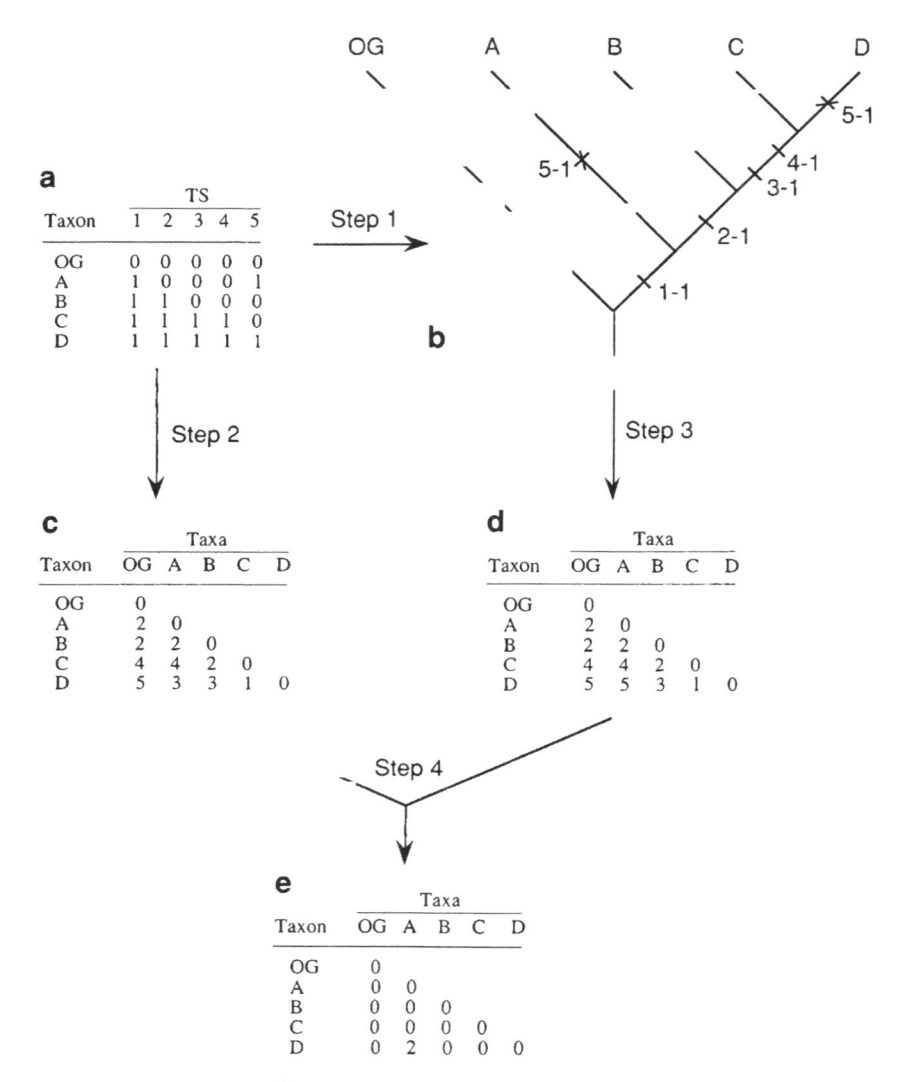

Fig. 5.2.—Matrices and tree illustrating F-ratio calculations (Example 5.2). **a.** Data matrix. **b.** Tree. **c.** Phenetic distance matrix. **d.** Patristic distance matrix. **e.** Difference matrix.

Tree Summaries Exercises

EXERCISE 5.1.—Using the data matrix in Table 5.3, derive the tree and calculate the tree length, CI, R, and RC.

Table 5.3.—Data matrix for calculating summary statistics (Exercise 5.1).

Taxon	Transformation series							
	1	2	3	4	5	6	7	8
X (outgroup)	0	0	0	0	0	0	0	0
A	1	1	0	0	0	0	0	1
B	1	1	1	1	0	0	0	0
C	1	1	1	1	1	1	1	0

EXERCISE 5.2.—Using the matrix in Fig. 5.2a, calculate the tree length, CI, R, and RC for the tree in Fig. 5.2b.

EXERCISE 5.3.—Calculate the tree length, CI, R, and RC for the taxa and characters shown in Table 5.4 and Fig. 5.3.

Table 5.4.—Data matrix for calculation of tree length, CI, R, and RC from Fig. 5.3 (Exercise 5.3).

Taxon	Transformation series							
	1	2	3	4	5	6	7	8
OG	0	0	0	0	0	0	0	0
A	1	0	0	0	1	1	1	1
B	0	1	0	0	1	1	1	0
C	0	0	1	0	0	1	1	0
D	0	0	0	1	0	0	1	1

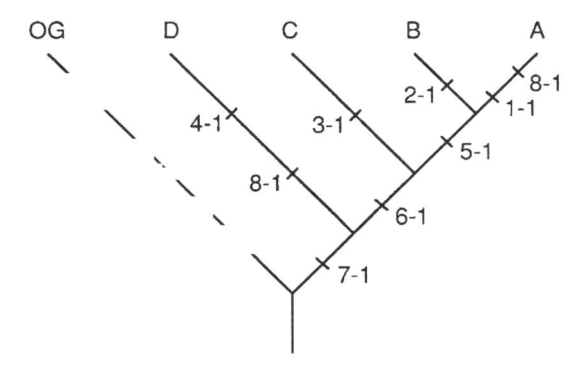

Fig. 5.3.—The tree used to calculate tree length, CI, R, and RC (Exercise 5.3).

EXERCISE 5.4.—Derive the tree and calculate the tree length, CI, R, and RC for the taxa and characters shown in Table 5.5.

Table 5.5.—Data matrix for calculation of tree length, CI, R, and RC (Exercise 5.4).

Taxon	\multicolumn{10}{c}{Transformation series}
	1	2	3	4	5	6	7	8	9	10
Outgroup	0	0	0	0	0	0	0	0	0	0
M	1	1	0	0	1	1	1	1	0	0
N	1	1	1	1	1	1	1	0	1	0
O	1	1	1	1	0	0	0	0	0	1

EXERCISE 5.5.—Derive the tree and calculate the tree length, CI, R, and RC for the taxa and characters shown in Table 5.6.

Table 5.6.—Data matrix for calculation of tree length, CI, R, and RC (Exercise 5.5).

Taxon	1	2	3	4	5	6	7	8	9	10	11
OG	0	0	0	0	0	0	0	0	0	0	0
A	0	0	0	0	0	0	0	0	0	0	1
B	1	1	1	0	0	1	1	0	0	0	0
C	1	1	1	0	0	1	1	0	0	1	1
D	1	1	1	1	1	1	0	1	1	0	0
E	1	0	1	1	1	0	0	1	1	0	0
F	1	1	1	0	0	0	0	1	1	0	1

CONSENSUS TECHNIQUES

One easy way of dealing with a set of topologically different but equally parsimonious trees or a set of topologically different trees derived from different data sets is to combine their information in some manner. **Consensus trees** are trees that combine the information, or knowledge claims, about grouping contained in two different trees into a single tree. As such, consensus trees must be used very carefully. It would be tempting, for example, to decide that a consensus tree, containing all the information from two different but equally parsimonious trees, gives you the best estimate of the phylogenetic relationships among groups. However, this is not true. In this section, we will cover some of the basic consensus techniques and give you some guidelines for using them. Like the Wagner algorithm section

TREE COMPARISONS

of Chapter 4, we do not expect that you will use the techniques presented here to generate consensus trees because the computational effort for even simple trees is rather great. Instead, these exercises are designed to give you a concept of what these trees do to groups elucidated by parsimony. We will cover three kinds of trees that can be generated by different consensus methods. Each can be used in different situations to answer different questions.

Strict Consensus Trees

Strict consensus trees (Sokal and Rohlf, 1981) contain only those monophyletic groups that are common to all competing trees. The ellipses in a Venn diagram can represent these groups. (We shall make extensive use of Venn diagrams in this and following sections, so you might want to look back at Chapter 1 for a quick review, then try the Venn diagram Quick Quiz.)

Example 5.4.—A strict consensus tree.

The two trees shown in Fig. 5.4 have different knowledge claims about the relationships of taxa C, D, and E. These trees are logically inconsistent, but they do contain some common knowledge claims.

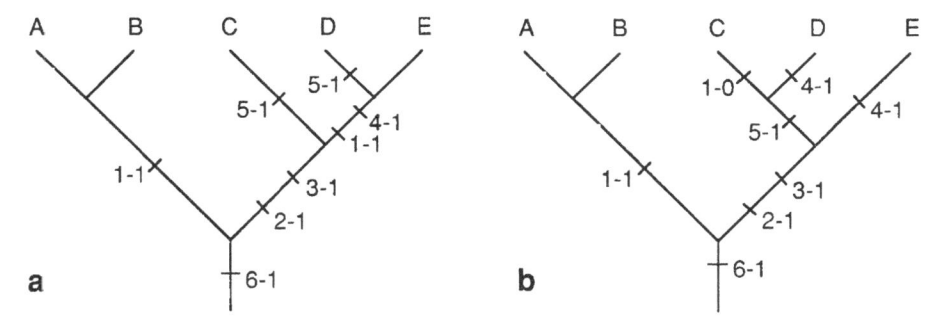

Fig. 5.4.—Two hypotheses of the relationships among taxa A–E (Example 5.4).

We construct a strict consensus tree for these alternative trees in the following manner.

1. Draw Venn diagrams for each tree (Fig. 5.5a, b).
2. Combine the Venn diagrams into a single diagram (Fig. 5.5c). It should come as no surprise that the groupings intersect.
3. Erase all intersecting ellipses (Fig. 5.5d). The result is a **consensus Venn diagram**.
4. Translate the consensus Venn diagram into a strict consensus tree (Fig. 5.6).

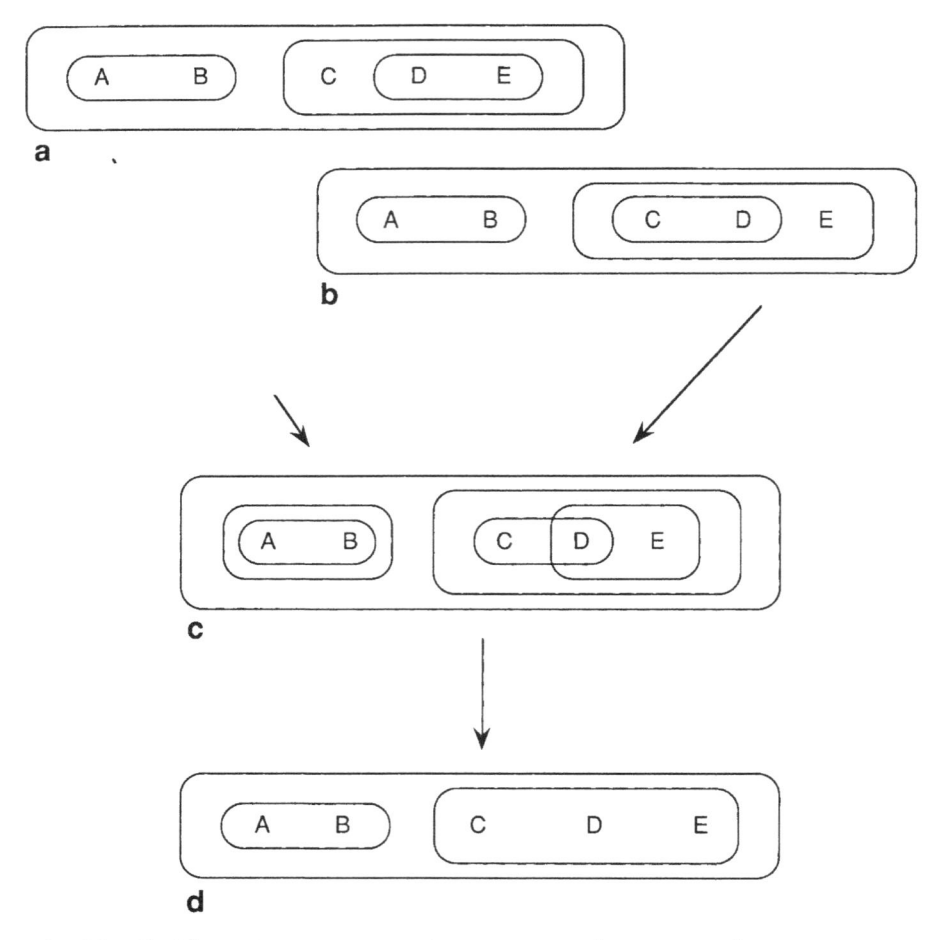

Fig. 5.5.—Obtaining a strict consensus Venn diagram. **a, b.** Venn diagrams of the original trees in Fig. 5.4. **c.** Combined Venn diagrams. **d.** Strict consensus Venn diagram.

EXERCISE 5.6.—Using the trees shown in Fig. 5.7, construct a strict consensus tree using Venn diagrams.

Example 5.5.—Another strict consensus tree.

Our first example concerned trees that contained logical inconsistencies. Strict consensus trees, however, also deal with different trees, or parts of trees that do not have logically inconsistent topologies. In such cases, the strict consensus tree would be the tree of lowest resolution. Consider the trees in Fig. 5.8a–c. If we employed Venn diagrams, we would see that these trees are logically consistent with each other. The strict consensus tree (Fig. 5.8d)

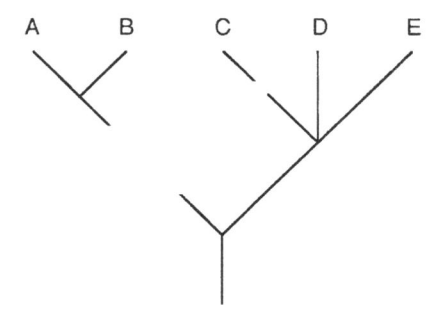

Fig. 5.6.—A strict consensus tree of the taxa A–E derived from Fig. 5.5d.

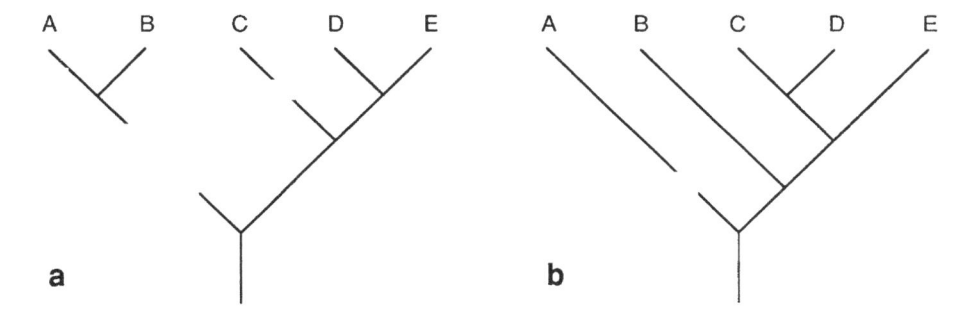

Fig. 5.7.—Phylogenetic trees for construction of strict consensus tree (Exercise 5.6).

represents the "lowest common denominator" concerning the knowledge claims of the three trees. For example, in moving A to the root node, we cannot maintain B and C as a monophyletic group, even though two of the three trees (5.8a, b) contain this grouping. Why? Because the third tree (5.8c) makes the claim that A, B, and C form a monophyletic group.

Adams Trees

Adams consensus trees are designed to give the highest "resolution" possible between two or more trees. When these trees are logically inconsistent, the taxa responsible for the conflict are relocated. Thus, Adams consensus trees do not necessarily reflect monophyletic groups that are supported by the original data matrix. A detailed description of Adams trees can be found in Adams (1972). He presents two cases, trees with labeled nodes (e.g., specified ancestors) and trees with unlabeled nodes (trees of common ancestry). We will deal only with the second case, trees with unlabeled nodes.

Adams techniques are built around several steps. The investigator partitions the taxa into sets across each of the trees, beginning with the root node. This partitioning groups taxa into sets based on their connection via a branch to the root node. For example, the partition sets

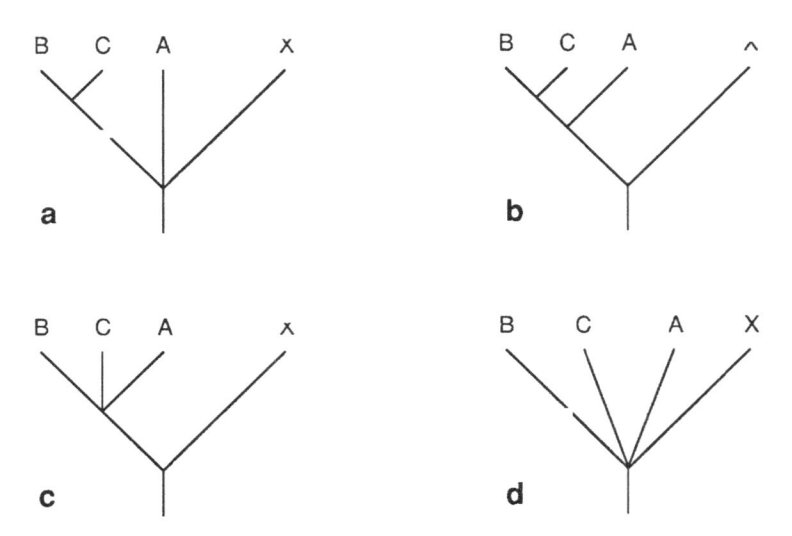

Fig. 5.8.—Three trees (a–c) and their strict consensus tree (d).

for Fig. 5.8b are {X} and {ABC}, the sets of taxa connected to the root via a branch. (Note that {ABC} is not subdivided into {A} and {BC} at this time. Such a subdivision would happen only in the next round when the node common to these taxa alone is considered.) The investigator then compares these sets across all trees to find partition products, which are the sets (=taxa and/or monophyletic groups of taxa) common to each of the original trees. For example, the tree in Fig. 5.8a has the partition sets {BC}, {A}, and {X}. The partition products of the trees 5.8a and 5.8b are [X], [A], and [BC]. The set {ABC} is not a product because it does not appear in the partition sets common to both trees. Note that single taxa as well as sets of taxa can be products. However, single taxa cannot be further partitioned, and sets with only two taxa are automatically partitioned into only the two single taxa. Note that nonempty partition products require at least one taxon to be part of the partition sets compared between trees. Finally, the method works from the root of the original trees outwards, identifying partition sets until the terminal taxa are reached. Any taxa partitioned out are not considered at higher nodes even if they are parts of monophyletic groups in one or more of the original trees.

Example 5.6.—For this example, we will use the phylogenetic trees shown in Fig. 5.8a–c. Because the trees contain only four taxa, we examine only two nodes; the second node we examine contains only two taxa.

1. Partition the taxa into partition sets connected to the root node. This yields the following partition sets for each tree.

Tree 5.8a (3 partition sets): {X}, {A}, {BC}
Tree 5.8b (2 partition sets): {X}, {ABC}
Tree 5.8c (2 partition sets): {X}, {ABC}

2. Determine the partition products by finding the intersections (common set elements) of each partition set that yields a nonempty partition product. The simplest of these partition products is the intersection of the sets {X}, {X}, and {X}, yielding [X] as the partition product. Another is formed from the intersections of {A}, {ABC}, and {ABC}, yielding [A] as the partition product. The final product is formed by the intersections of {BC}, {ABC}, and {ABC}, yielding [BC] as the partition product.

3. From the root, proceed to the next node that contains more than one taxon and repeat the partitioning and determine the partition products as in steps 1 and 2. Repeat up the branch or along sister branches until all terminal taxa are partitioned. In this example, there is only one final node that needs to be considered, the node leading to {BC} in tree 5.8a and to {ABC} in trees 5.8b and 5.8c. Note that {A} has already been partitioned, so the only products of partition are [BC] for each of the three trees.

4. Form the Adams consensus tree from the sets determined by partitioning, beginning at the root and using the partition products as sets. Figure 5.9a shows the Adams consensus tree derived from step 4. The first partition products, [A], [X], and [BC], are joined at the root node. The set {BC} is automatically partitioned (there are only two terminal taxa), yielding the dichotomy (Fig. 5.9b).

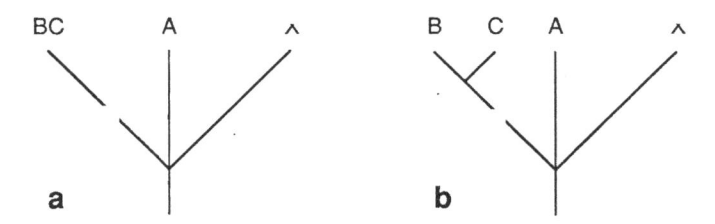

Fig. 5.9.—Adams consensus tree for Example 5.6. a. Result of adding the first partition products. b. Final Adams consensus tree.

Example 5.7.—For this example, we will use the trees shown in Fig. 5.10a–c. These trees contain more taxa and nodes and will allow us to form some additional partition products that are not trivial (i.e., that contain more than one or two taxa).

1. Form the first partition sets for the three trees. Note that the partition sets of trees 5.10a and 5.10c are exactly the same, although the hypotheses of common ancestry *within* these groups are different.

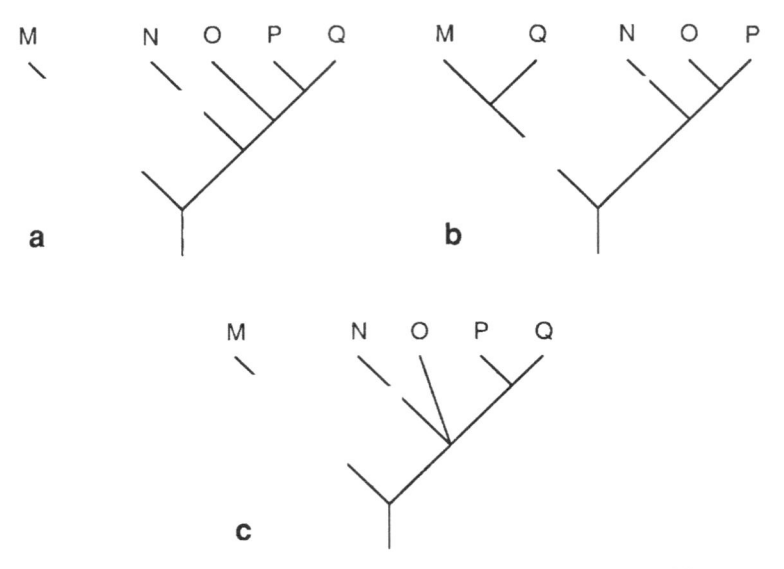

Fig. 5.10.—Trees for construction of Adams consensus tree (Example 5.7).

Tree 5.10a (2 partitions): {M}, {NOPQ}
Tree 5.10b (2 partitions): {MQ}, {NOP}
Tree 5.10c (2 partitions): {M}, {NOPQ}

2. Determine the partition products.

Partition product 1: {M}, {MQ}, {M} = [M]
Partition product 2: {NOPQ}, {NOP}, {NOPQ} = [NOP]
Partition product 3: {NOPQ}, {MQ}, {NOPQ} = [Q]

Note that the partition set {NOPQ} of both trees 5.10a and 5.10c have been used twice because one element (Q) appears in a different section of Fig. 5.10b and the other elements appear in a different set {NOP}. Also note that other combinations are possible, but they contain no interesting taxa. For example, the intersection of {M} from tree 5.10a and {NOPQ} from 5.10c yield an empty set.

3. Proceed to the next node up from the root of each tree and determine the partition sets, deleting any taxa that might have already been partitioned as single taxa.

Tree 5.10a (2 partitions): {N}, {OP}
Tree 5.10b (2 partitions): {N}, {OP}
Tree 5.10c (3 partitions): {N}, {O}, {P}

Note that Q does not appear in either tree 5.10a or 5.10c because it has already been partitioned. As we will see, Q will join the Adams tree at the root because it is a partition product of the root node.

4. Form the partition products.

TREE COMPARISONS

Partition product 1: {N}, {N}, {N} = [N]
Partition product 2: {OP}, {OP}, {O} = [O]
Partition product 3: {OP}, {OP}, {P} = [P]

5. All terminal taxa have been partitioned. Form the consensus tree, beginning from the root and connecting each product of the partition associated with each node (Fig. 5.11).

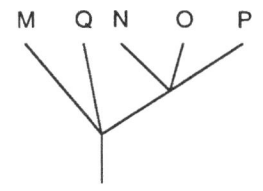

Fig. 5.11.—Final Adams consensus tree for Example 5.7.

EXERCISE 5.7.—Using the trees in Fig. 5.12, construct an Adams consensus tree.

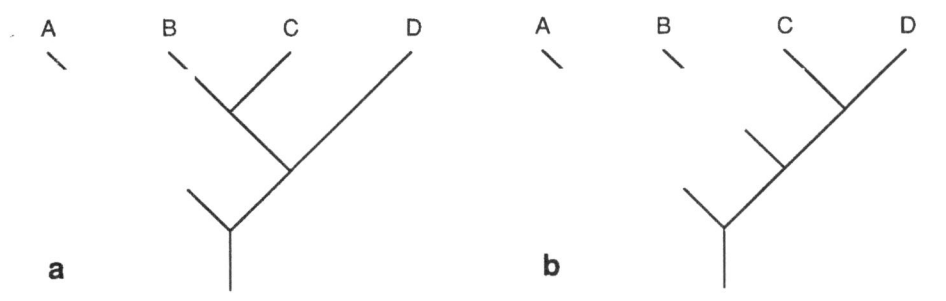

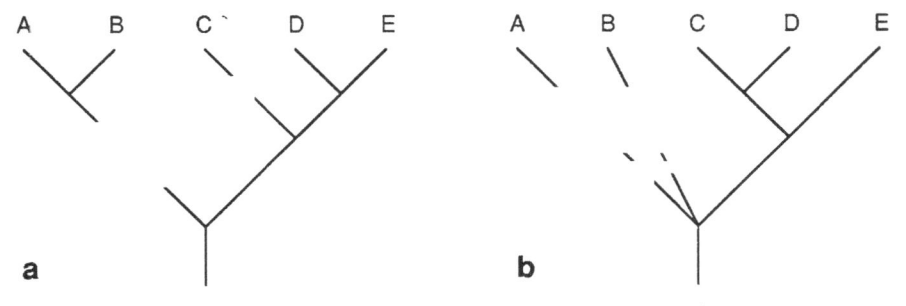

Fig. 5.12.—Trees for construction of Adams consensus tree (Exercise 5.7).

EXERCISE 5.8.—Using the trees in Fig. 5.13, construct an Adams consensus tree.

Fig. 5.13.—Trees for construction of Adams consensus tree (Exercise 5.8).

Majority Consensus Trees

Majority consensus trees operate on a "majority rule" basis (Margush and McMorris, 1981), therefore the consensus tree may be logically inconsistent with one or more of the original parsimony trees. If you have one tree that hypothesizes that A and B are more closely related to each other than either is to C (Fig. 5.14a) and two trees that hypothesize that B and C are more closely related to each other than either is to A (Fig. 5.14b, c), then the majority consensus tree will have the latter topology (Fig. 5.14d). If there is an "even vote," then the result is a polytomy. If the trees are logically consistent, then the most resolved tree is preferred (just as in the case of Adams consensus trees).

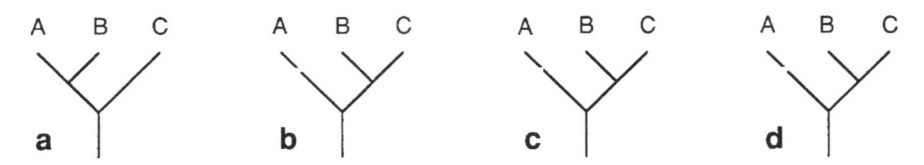

Fig. 5.14.—Three trees (a–c) and the majority consensus tree (d) for taxa A–C.

EXERCISE 5.9.—Use the principles presented above to construct a majority consensus tree for the trees shown in Fig. 5.15.

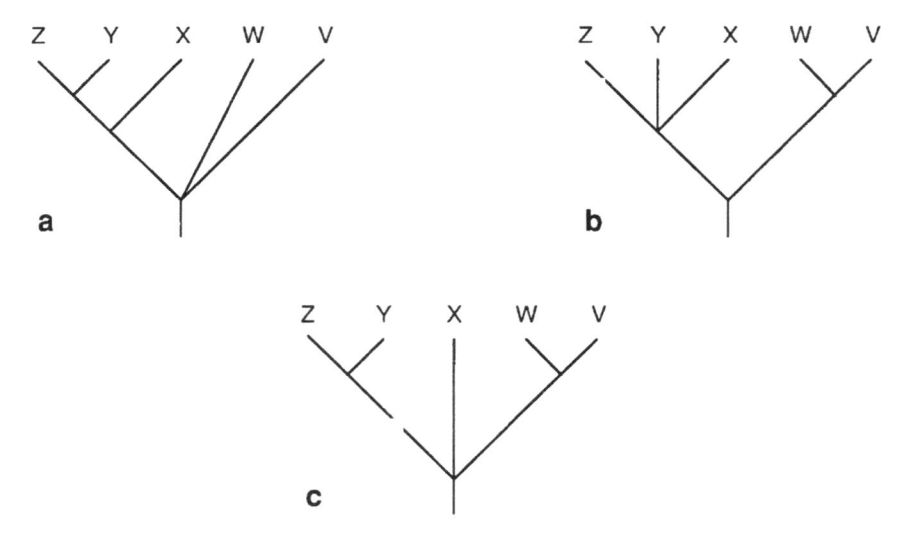

Fig. 5.15.—Trees for construction of majority consensus tree (Exercise 5.9).

EXERCISE 5.10.—Using the trees in Fig. 5.16, construct a strict consensus tree, an Adams consensus tree, and a majority consensus tree.

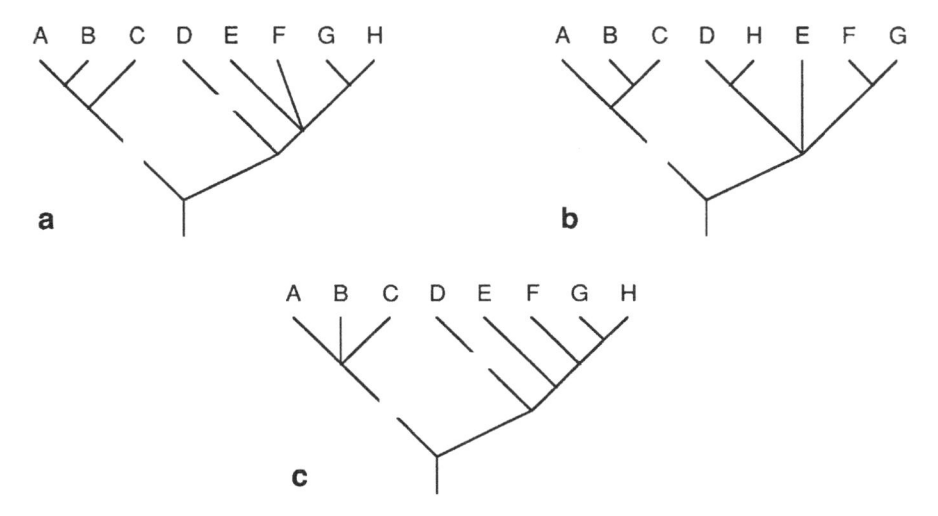

Fig. 5.16.—Trees for construction of strict, Adams, and majority consensus trees (Exercise 5.10).

CHAPTER NOTES AND REFERENCES

1. The use of CI to characterize the amount of homoplasy for a particular tree is controversial, and other indices have been suggested, including the homoplasy excess ratio (Archie, 1989) and the retention index (Farris 1989b). See also papers by Farris (1990) and Archie (1990) for various opinions regarding these measures and the use of CI.

2. Nelson (1979) is often credited as the source for strict consensus trees. However, he really described component analysis, which possesses qualities of clique analysis (Page, 1987, 1988, 1989). The resulting trees may fit the description of either strict or majority rule consensus trees or no particular consensus tree at all, depending on the number of trees involved. The use of consensus trees of all types has increased in the last few years, probably because they offer what appears to be a simple solution to the difficult problems associated with choosing among several equally parsimonious trees. However, we feel that consensus trees solve no such problems and their use in this manner has the effect of avoiding the difficult problems associated with equally parsimonious trees. Having many equally parsimonious trees may be the result of insufficient study (a lack of sufficient data) or chaotic evolution in the particular data sets studied, reflecting a complex history of reticulation or homoplastic evolution. You should be extremely careful when using consensus trees to investigate specific questions or to organize your data. A consensus tree

should not be presented as a phylogeny unless it is topologically identical with one or more of the most parsimonious trees.

3. Competing trees fall into three overlapping categories. 1) The competing trees may be equally parsimonious. 2) They may be equally parsimonious but have different topologies. 3) They may be close but not equal in terms of tree length. Equally parsimonious trees that have the same topology result from different interpretations of character evolution. Frequently, this involves equally parsimonious interpretations of parallelism and reversals. When different topologies are involved, the trees have different knowledge claims concerning the relationships of the taxa studied. Such trees may be logically consistent or logically inconsistent with each other. You can use the Adams consensus method to explore these differences. Loss of resolution occurs when there is an increase in the number of nodes with polytomies or when the number of branches involved in a polytomy increases. The taxon that causes a loss of resolution must have different placements on at least two trees. You can check the original trees just in these areas, or you can do pairwise comparisons of the original trees to identify the conflicts. Once these trees have been identified, you can find the taxa involved and the characters that are responsible for the conflict. You might also use the strict consensus method to determine what monophyletic groups are supported by all of the original trees. Finally, you might use the Adams method for a set of trees that are within certain tree-length values of each other. Funk (1985) found this approach useful in her study of hybrid species, when the differing placements of certain species helped spot potential hybrids and lead to hypotheses concerning their parental species. Table 5.7 summarizes some of the uses and potential problems with various kinds of consensus trees.

Table 5.7.—Uses and potential problems of the various kinds of consensus trees.

Kind of tree	Questions asked	Characteristics
Strict	1) What groups are always monophyletic?	1) Loss of resolution may be extreme. 2) Useful as a phylogeny only if topologically identical with one of the original parsimony trees.
Adams	1) What is the most highly resolved tree that will identify problem taxa? 2) Are these trees logically consistent?	1) "Strange" taxon placements not found in any original tree. 2) Useful as a phylogeny only if topologically identical with one of the original parsimony trees.
Majority	1) What is a summary of the competing trees where the dominant pattern prevails?	1) Most useful when there are very little conflicting data. 2) Useful as a phylogeny only if topologically identical with one of the original parsimony trees. 3) Ties can be avoided only if you start out with an odd number of trees (A. Kluge, pers. comm.).

CHAPTER 6

CLASSIFICATION

A **classification** is a groups-within-groups organization of taxa of organisms. This type of organization can be represented in many different ways. Most classifications take the form of Linnaean hierarchies, with the relative positions of groups and subgroups being tagged with a Linnaean rank (phylum, family, genus, etc.). Classifications can also be represented graphically, using a tree structure or a Venn diagram (see Chapter 1).

Nothing marks the phylogenetic system as different from competing systems so much as the issue of classification. Many investigators are drawn to the advantages of using phylogenetic techniques to infer common ancestry relationships, but they seem to balk at excluding paraphyletic groups. Much acrimony is produced in the name of "tradition," "common sense," or other emotional side issues, rather than focusing in on the one central issue: should biological classifications be based on phylogeny? We answer "yes" because we are evolutionary biologists. Actually, phylogenetic systematists operate under only two basic principles. First, a classification must be consistent with the phylogeny on which it is based. Second, a classification should be fully informative regarding the common ancestry relationships of the groups classified. Further, we recommend that while implementing these principles, you make every attempt to alter the current classification as little as possible.

The principle of consistency can be met if the investigator includes only monophyletic groups. The principle of information content can be approached by several means. The conventions we outline in this chapter will allow you to change existing classifications minimally to bring them "in line" with current hypotheses of genealogical relationships. The principle of minimum change represents a conservative approach that provides for historical continuity of classification in the change from preevolutionary concepts (such as distinctiveness and identity) to evolutionary concepts (common ancestry relationships).

Often, taxonomists are categorized as either "splitters" or "lumpers." Actually, phylogeneticists are neither. By following the principles of monophyly and maximum information content, we seek to establish classifications that reflect natural groups. Sometimes this requires breaking up a paraphyletic or polyphyletic group into smaller groups. At other times these goals are met by combining smaller groups into more inclusive groups. The categories of "splitters" and "lumpers" belong to the past when authority was more important than data.

The topics covered in this chapter are 1) how to evaluate existing classifications relative to new ideas about the common ancestry relationships of the organisms studied and 2) how to construct classifications using conventions designed to conserve as much of the old taxonomies as possible.

Evaluation of Existing Classifications

Once the investigator has analyzed her data and arrived at a phylogenetic tree (a hypothesis of common ancestry relationships), she will wish to compare her results with the ideas found in previous studies. Most of the ideas of relationship that exist in the literature are embodied in classifications. These classifications have connotations regarding the relationships that the previous investigator felt were important to the taxonomy of the group. The ideas embodied in the classification may represent the intuition of the previous investigator or they may refer very specifically to some evolutionary principle the investigator had in mind. Whatever the ideas held by the original taxonomist, it is up to our investigator to compare her hypothesis with the existing classification to determine how that classification should be changed to bring it in line with the phylogeny. This is accomplished by comparing the structure of the classification with the structure of the phylogenetic tree. If the existing taxonomy is logically consistent with the phylogeny, then the first criterion is met. If it is fully informative about the phylogeny, then the second criterion is met. Because logical consistency is the most basic requirement, we will cover it first.

Logical Consistency

As outlined by Hull (1964), logical consistency exists between a classification and a phylogeny if, and only if, at least one phylogeny can be derived from the classification that is, itself, the phylogeny. You might get the idea that it may be possible to derive more than one phylogeny from a classification. This is true for some classifications but not for others. However, a more basic question must be addressed. How do we actually go about comparing a phylogeny and a classification? The answer is fairly straightforward. The phylogeny exists in tree form and the classification exists in the form of a groups-within-groups hierarchy. We therefore must convert the hierarchy of the classification into tree form. Once in tree form, we can make a side-by-side comparison.

Example 6.1.—Converting a classification to tree form.

The following classification is of the family Goodeidae, North American and Middle American killifishes related to the guppies and swordtails and commonly found in aquarium stores (Parenti, 1981).

Family Goodeidae
    Subfamily Empetrichthyinae
        Genus *Empetrichthys*
        Genus *Crenichthys*
    Subfamily Goodeinae
        (several genera)

We convert the classification to tree form using the groupings inherent in the classification; for example, there are two subfamilies within the family. The classification in tree form looks like Fig. 6.1. (We have included the names of the groups and subgroups to show the direct connection between this tree and the original classification. It is not necessary to do so once you have the idea firmly in mind.)

Empetrichthys   *Crenichthys*   Goodeinae

Empetrichyinae

Goodeidae

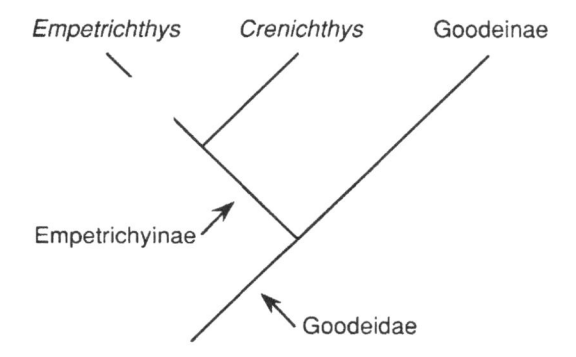

Fig. 6.1.—Parenti's (1981) classification of Goodeidae in tree form.

There are several important things about this classification in tree form. First, it exactly reflects the groups-within-groups organization of the ranked classification. Internodes take the place of higher group names. Second, it is entirely dichotomous. This means that no additional diagrams can be derived from it. Classifications that are fully ranked and thus have tree structures that are entirely dichotomous have no derivatives. Last, **it is not a phylogeny**. We cannot stress this last point enough. The classification might be based on phenetic principles. It might be arbitrary. It might be intuitive. Or, it might be phylogenetic. To know which, we would have to read the paper. None of us is an expert in goodeid relationships, but it turns out that Parenti (1981) also performed a phylogenetic analysis, so we can take her hypothesis as a basis for comparison (Fig. 6.2).

Now, we can place the classification beside the phylogeny. Note that they are entirely identical in their topologies. No claims of group-within-group relationships are different from the claims of common ancestry. Let us imagine that Parenti had chosen to classify her groups in a different way.

Family Goodeidae
    Subfamily Empetrichthyinae
      Genus *Empetrichthys*
    Subfamily Crenichthyinae
      Genus *Crenichthys*
    Subfamily Goodeinae
      (several genera)

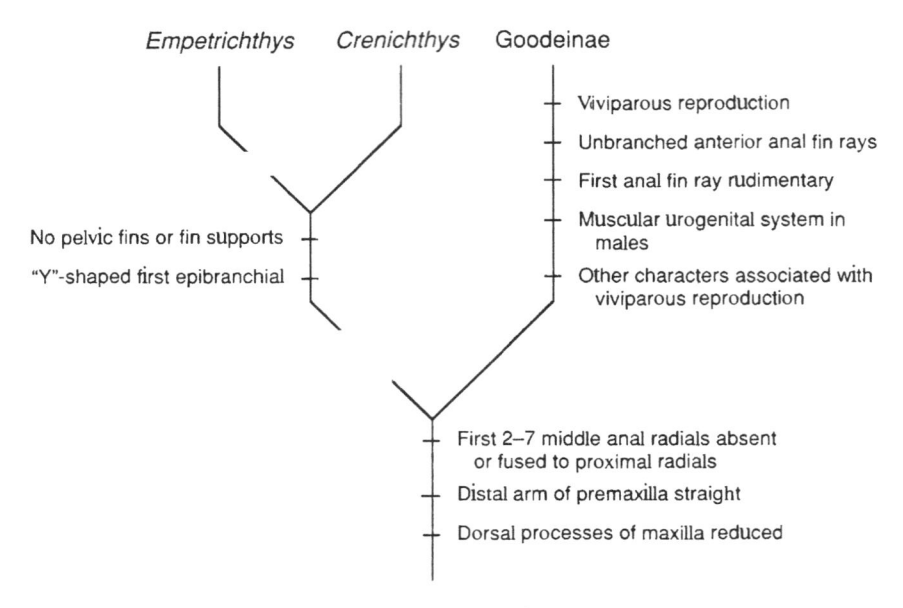

```
                Empetrichthys   Crenichthys   Goodeinae
                                               ├── Viviparous reproduction
                                               ├── Unbranched anterior anal fin rays
                                               ├── First anal fin ray rudimentary
   No pelvic fins or fin supports ┤            ├── Muscular urogenital system in
                                               │   males
   "Y"-shaped first epibranchial  ┤            ├── Other characters associated with
                                               │   viviparous reproduction

                        ├── First 2–7 middle anal radials absent
                        │   or fused to proximal radials
                        ├── Distal arm of premaxilla straight
                        ├── Dorsal processes of maxilla reduced
```

Fig. 6.2.—A phylogenetic tree of goodeid relationships (from Parenti, 1981).

Conversion of the classification to tree form would produce Fig. 6.3.

We now have a situation in which there is a trichotomy among the three subgroups. Whenever there is a polytomy, we can derive additional hypotheses about possible subgroupings. These hypotheses represent all of the possible resolutions of the classification tree, given its structure. That is, we can derive possible resolutions that are consistent with the original classification. Following the principle of logical consistency, if one of these

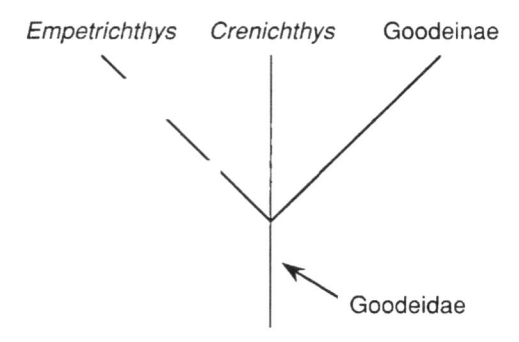

```
        Empetrichthys   Crenichthys   Goodeinae
                  \         |         /
                   \        |        /
                    \       |       /
                     \      |      /
                      \     |     /
                       \    |    /
                        \   |   /
                         \  |  /
                          \ | /
                           \|/
                            | ← Goodeidae
                            |
```

Fig. 6.3.—An alternative classification of goodeids in tree form.

CLASSIFICATION 95

derivatives is the phylogeny, then the classification is logically consistent with the phylogeny even if it is not fully informative about the relationships hypothesized in the phylogeny. The number of possible trees can be determined mathematically. For the trichotomy in Fig. 6.3, there are three dichotomous resolutions (Fig. 6.4a–c). Note that one of them (Fig. 6.4b) is topologically identical to the phylogeny. Therefore, even though the branching topology of the classification (Fig. 6.3) is not the same as the topology of the phylogenetic hypothesis (Fig. 6.2), the classification is logically consistent with the phylogeny because one possible derivation of the classification is topographically identical to the phylogeny.

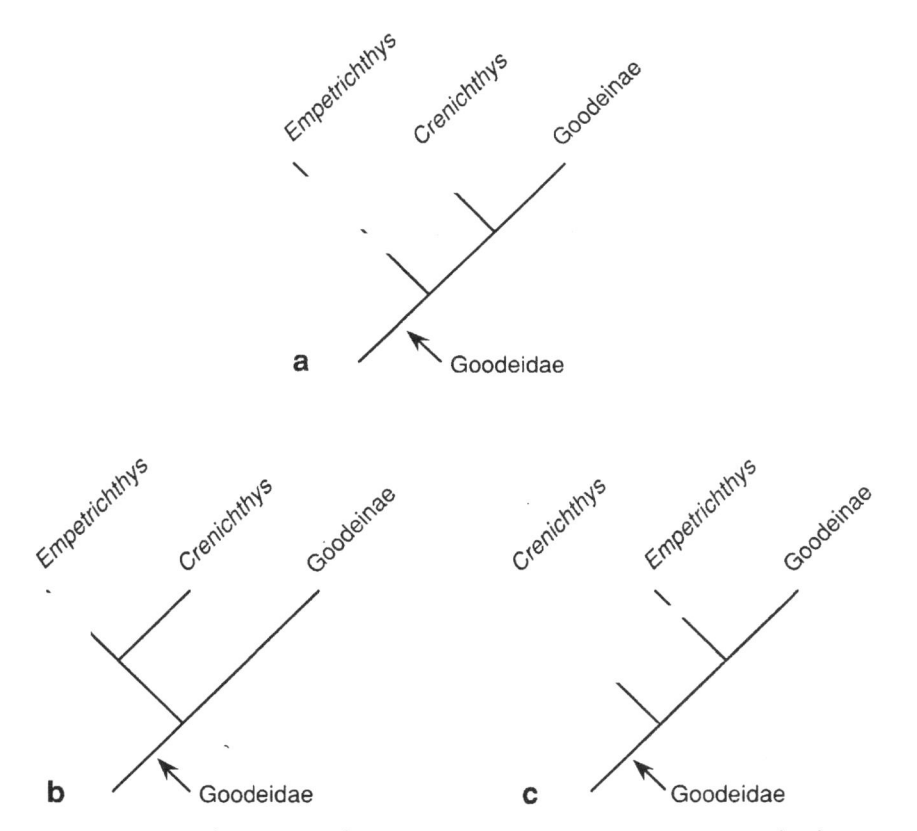

Fig. 6.4.—Three possible phylogenetic trees of goodeids derived from the relationships implied by the classification in Fig. 6.3.

Working out derivative classifications in tree form is easy when the number of possible derivatives is small. The method becomes cumbersome, however, when the number of possible derivative trees is large. Fortunately, there is an alternative method for checking the logical consistency of a classification with a particular phylogeny, the use of Venn diagrams.

In Fig. 6.5, we have converted the phylogeny (Fig. 6.5a), the first classification (Fig. 6.5b), and the second classification (Fig. 6.5c) into Venn diagrams. We then "layer" all three diagrams and check for overlap between ellipses, which is equivalent to checking for overlapping sets in set theory. If there are no overlaps, the classification(s) is logically consistent; if there is overlap, the classification(s) is logically inconsistent. In this case there is no overlap, so both classifications are logically consistent with the phylogeny (Fig. 6.5d).

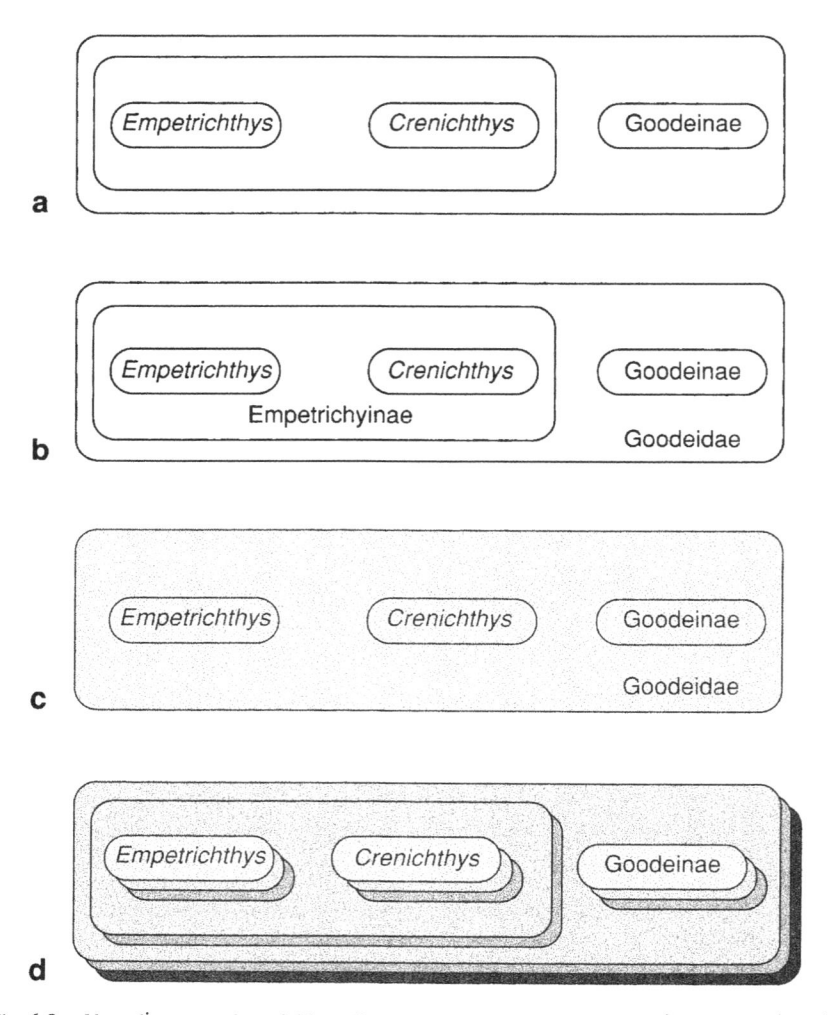

Fig. 6.5.—Venn diagrams of goodeids. a. The goodeid phylogeny. b, c. Two different classifications. d. The result of combining a, b, and c into a single diagram.

You can easily see that a classification can be logically consistent with a phylogeny without being fully informative about the common ancestry relationships implied in the phylogeny. In fact, the only way to have a classification that is logically inconsistent with the phylogeny is to have a classification for which we can derive no tree that is the original phylogeny. More formally, a classification is logically inconsistent with a phylogeny if no derivative of that classification is the original phylogeny. In terms of our Venn diagrams, a classification is logically inconsistent with the phylogeny if its Venn diagrams have overlapping ellipses. In other words, they are logically inconsistent if they violate the inclusion/exclusion rule.

We may be making it sound like most classifications are logically consistent with the phylogeny you are likely to generate with your analysis. Nothing could be further from the truth. You will find that most classifications are logically inconsistent with your hypotheses of common ancestry. Why? Because most existing classifications contain paraphyletic and even polyphyletic groups. Let's look at the effects of the inclusion of a paraphyletic group in the correspondence of classifications to phylogenies.

Example 6.2.—The very distinctive *Cus*.

Investigator Smith has performed a phylogenetic analysis on the genera comprising the family Cidae. She has arrived at the phylogenetic hypothesis shown in Fig. 6.6. This family was well known to previous investigators. What struck these investigators was how different members of the genus *Cus* were from other members of the family. This distinctiveness was embodied in the traditional classification.

Family Cidae
  Subfamily Ainae
    Genus *Aus*
    Genus *Bus*
  Subfamily Cinae
    Genus *Cus*

Smith wants to know if the current classification is logically consistent with her phylogenetic hypothesis. To do so, she must perform the following steps.

1. She prepares a Venn diagram for the classification and another Venn diagram for the phylogeny.
2. She layers the Venn diagram of the classification over the Venn diagram of the phylogeny.
3. If ellipses do not overlap, then she knows that the classification is logically consistent with the phylogeny. If one or more ellipses overlap, then the classification is logically inconsistent with the phylogeny.

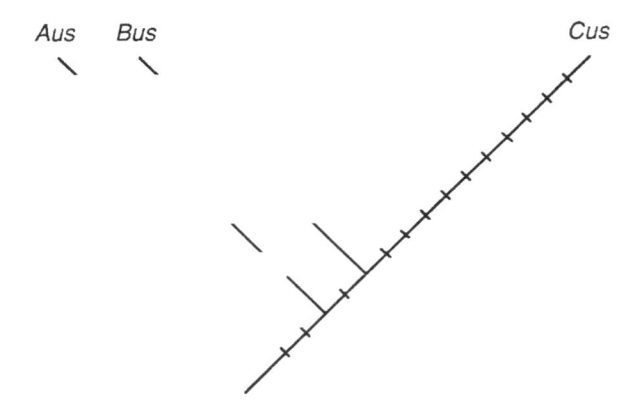

Fig. 6.6.—The phylogeny of the family Cidae (Example 6 2). The phylogeny and the relative amounts of change along each branch are taken as "true."

Therefore, Smith does the following for the Cidae.

1. The classification is converted to a classification tree (Fig 6 7a) and then into a Venn diagram (Fig. 6.7c). (As you gain experience, you can draw the Venn diagram directly from the classification.) The phylogeny (Fig. 6.7b) is then converted into a Venn diagram (Fig. 6.7d).

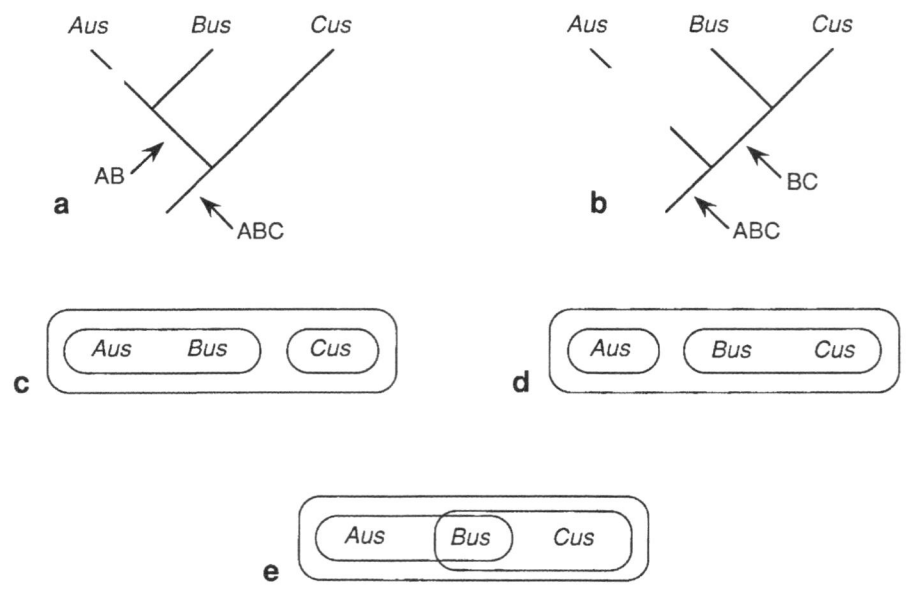

Fig. 6.7.—The Cidae classification. a. Tree form. b. The phylogeny. c. Venn diagram of a. d. Venn diagram of b. e. Venn diagram combining c and d.

2. The Venn diagrams are layered (Fig. 6.7e).
3. The Venn diagram of the phylogeny (Fig. 6.7d) and the Venn diagram of the classification (Fig. 6.7c) overlap (Fig. 6.7e); *Bus* is a member of the group BC in the Venn diagram of the phylogeny, and it is a member of the group AB in the Venn diagram of the classification. Therefore, the classification is logically inconsistent with the phylogeny.

Smith, realizing that she cannot tolerate a classification that is logically inconsistent with the evolution of the Cidae, creates the following classification.

Family Cidae
  Subfamily Ainae
    Genus *Aus*
  Subfamily Cinae
    Genus *Bus*
    Genus *Cus*

Determining the Number of Derivative Classifications

Although Venn diagrams are the most direct route to determining whether a classification is logically consistent with a phylogeny, you might also wish to calculate the number of derivative classification trees for a particular classification. The number of alternative classifications that can be derived from a particular classification is directly related to the number of polytomies in the classification's branching structure. Felsenstein (1977) presents tables to determine the number of tree topologies that can be derived from a basic tree with multifurcations (polytomies). We have reproduced parts of one of these tables as Table 6.1. Note that the numbers refer only to the number of terminal taxa. Other tables must be consulted if ancestors are included.

Example 6.3.—Classification of the hypothetical Xaceae.

The Xaceae is classified into three major subgroups, as shown in the classification tree in

Table 6.1.—The total number of possible derivative trees for polytomies of *n* branches. Internodes cannot be occupied by "ancestors" (from Felsenstein, 1977).

	All trees	Dichotomous trees
3	4	3
4	26	15
5	236	105
6	2752	945
7	39,208	10,395
8	660,032	135,135
9	12,818,912	2,027,025
10	282,137,824	34,459,425

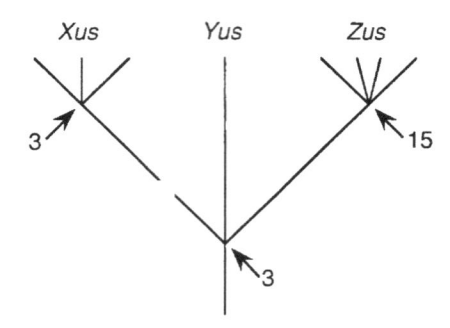

Fig. 6.8.—The classification tree of the Xaceae (Example 6.3).

Fig. 6.8. We calculate the number of trees that can be derived from this classification in the following manner.

1. Determine if the phylogeny to be compared is dichotomous or contains polytomies. In this example, we assume that it is a dichotomous phylogeny.
2. Select the appropriate column in Table 6.1. In this case, we use the column on the right (dichotomous trees).
3. Determine the number of branches for each polytomy.
4. Using Table 6.1, find the number of trees possible for each polytomy. Multiply all of the values obtained in step 3 together to obtain the total number of derivative classification trees:

$$(3)(3)(15) = 135.$$

Thus, there are 135 possible dichotomous trees that can be derived from our classification of Xaceae.

Classification Evaluation Exercises

Each exercise consists of one or more classifications and a phylogenetic tree. You are asked to 1) convert the classification into tree form, 2) state the number of possible derivative trees that can be obtained, and 3) evaluate the classification in terms of its consistency with the phylogeny.

EXERCISE 6.1.—Phylogeny of the Recent tetrapod vertebrates.

For classifications, see Fig. 6.9. For the phylogenetic tree, see Fig. 6.10.

CLASSIFICATION

a Lissamphibia
   Reptilia
     Chelonia
     Lepidosauria
     Crocodylia
   Aves
   Mammalia

b Lissamphibia
   Mammalia
   Chelonia
   Lepidosauria
   Crocodylia
   Aves

c Tetrapoda
   Lissamphibia
   Amniota
     Mammalia
     Reptilia
       Chelonia
       Sauria
         Lepidosauria
         Archosauria
           Crocodylia
           Aves

Fig. 6.9.—Three classifications of Recent tetrapod vertebrates (Exercise 6.1).

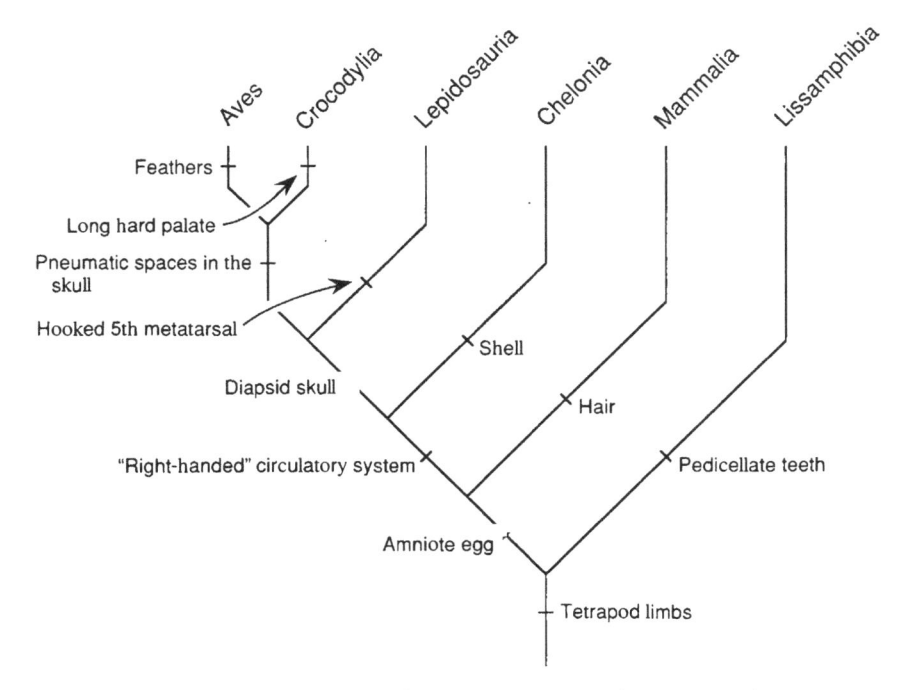

Fig. 6.10.—A phylogenetic hypothesis of Recent tetrapod relationships (Exercise 6.1).

EXERCISE 6.2.—Phylogeny of the land plants.

Use the following classification.

Division Bryophyta
    Class Anthoceropsida
    Class Marchantiopsida
    Class Bryopsida
Division Tracheophyta
    Subdivision Psilotophytina
    Subdivision Lycopodophytina
    Subdivision Sphenophytina
    Subdivision Pteridophytina
    Subdivision Spermatophytina
        Class Cycadopsida
        Class Pinopsida
        Class Ginkgopsida
        Class Gnetopsida
        Class Angiospermopsida

For the phylogenetic tree, see Fig. 6.11.

CONSTRUCTING PHYLOGENETIC CLASSIFICATIONS

There are two basic ways to construct phylogenetic classifications. First, one can consistently place sister groups in the classification at the same rank or rank equivalent. In such a classification, every hypothesized monophyletic group is named. If this manner of classifying is adopted, rank within a restricted part of the classification denotes relative time of origin. Second, one can adopt a set of conventions designed to reflect the branching sequence exactly but not require that every monophyletic group be named.

We consider it beyond the scope of this workbook to detail the controversies surrounding such topics as the suitability of Linnaean ranks versus indentation or numerical prefixes for constructing classifications (reviewed in Wiley, 1981a). Instead, we will briefly review some of the basics of phylogenetic classification and provide a summary of some of the conventions you might use in constructing your classifications. This will be followed by some exercises designed to demonstrate when certain conventions might be used.

Rules of Phylogenetic Classifications

Rule 1.—Only monophyletic groups will be formally classified.
Rule 2.—All classifications will be logically consistent with the phylogenetic hypothesis accepted by the investigator.
Rule 3.—Regardless of the conventions used, each classification must be capable of expressing the sister group relationships among the taxa classified.

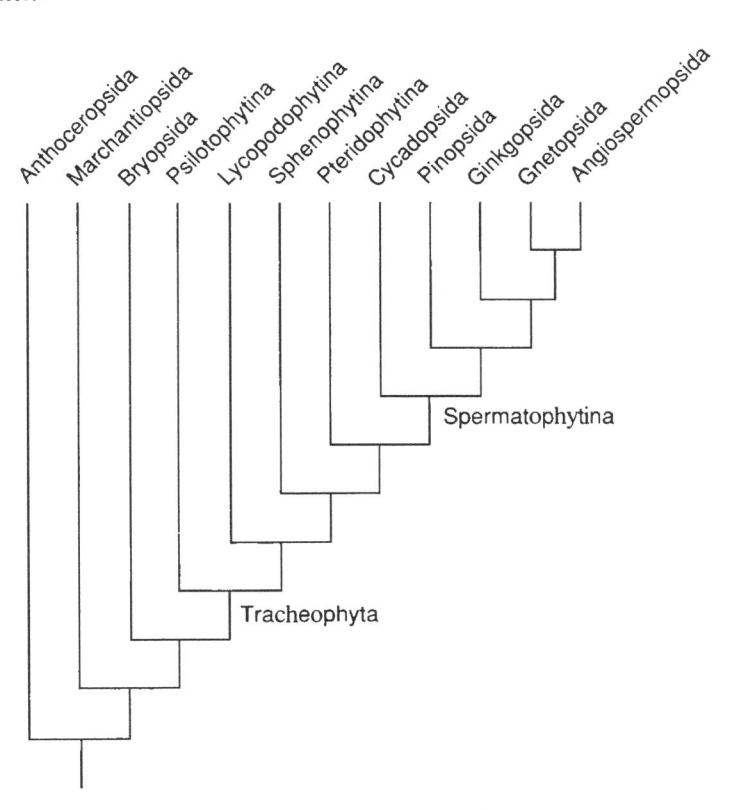

Fig. 6.11.—A phylogenetic hypothesis of selected land plant relationships (Exercise 6.2). (Name endings for terminal taxa correspond to classification for convenience.)

Conventions

We list the conventions used by Wiley (1981a) in his "annotated Linnaean hierarchy" system.

Convention 1.—The Linnaean system of ranks will be used.

This is one of the controversial conventions. There have been many attempts to substitute other means of subordinating taxa, and some of these attempts are mentioned at the end of this chapter. None are, in our opinion, satisfactory. However, the Linnaean system of ranks is, itself, a convention. There is no biological or scientific imperative for using Linnaean ranks.

Convention 2.—Minimum taxonomic decisions will be made to construct a classification or to modify existing classifications.

This convention can be met when making a new classification by rejecting redundant taxon names. For example, if a family has only one genus, do not coin a subfamily name. The subfamily name would be redundant with the family name. Of course, the family name is redundant, in terms of its diagnosis, with the genus. But, the family name serves another purpose, to place the genus within the context of its phylogenetic relationships with other genera. When converting existing classifications, make every effort to retain well known taxon names at their traditional ranks. This may or may not be possible.

Convention 3.—Taxa forming an asymmetrical part of a phylogenetic tree may be placed at the same rank and sequenced in their order of branching (Nelson, 1972). When such a list is encountered, the sequence of the list denotes the sequence of the branching.

This convention can be used by an investigator who decides that he will not name every monophyletic group but still wishes to preserve Rule 3. This convention is frequently termed the "listing convention" or "sequencing convention" and was first proposed by Nelson (1972) (Fig. 6.12).

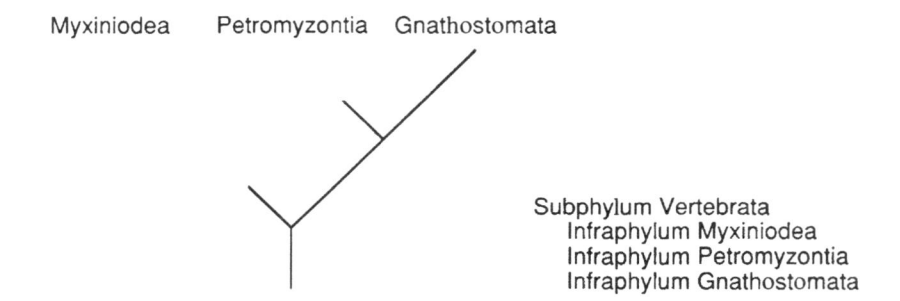

Fig. 6.12.—A hypothesis of the relationships and a classification of three vertebrate groups, illustrating the sequencing convention (Convention 3).

Convention 4.—Taxa whose relationships are polytomous will be placed *sedis mutabilis* at the same rank (Wiley, 1979).

This convention must be used if the sequence convention is used because you must be able to discriminate a list that translates into a series of dichotomies from a list that translates into a polytomy (Fig. 6.13).

Convention 5.—Monophyletic taxa of uncertain relationships will be placed *incertae sedis* at a level in the hierarchy where their relationships are known with some certainty.

This convention covers the situation where a small monophyletic group is thought to belong to, say, an order but cannot be placed in any suborder, family, etc., within that order.

CLASSIFICATION

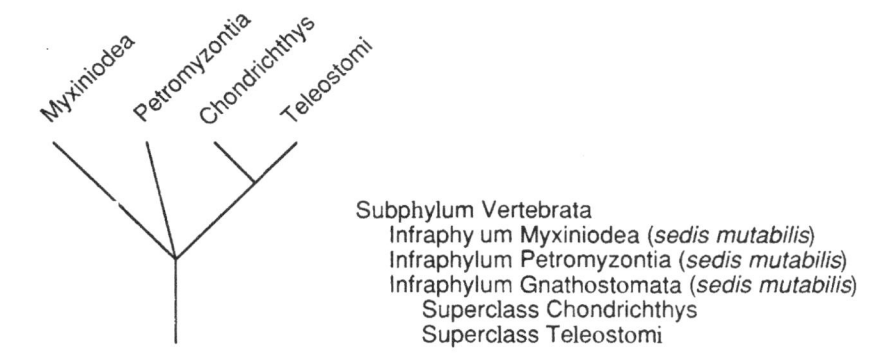

Subphylum Vertebrata
  Infraphylum Myxiniodea (*sedis mutabilis*)
  Infraphylum Petromyzontia (*sedis mutabilis*)
  Infraphylum Gnathostomata (*sedis mutabilis*)
  Superclass Chondrichthys
  Superclass Teleostomi

Fig. 6.13.—Another hypothesis of the relationships among vertebrates, with gnathostomes resolved into its two major subgroups, illustrating the *sedis mutabilis* convention (Convention 4).

Convention 6.—A group whose qualities are not known may be included in a phylogenetic classification if it is treated as *incertae sedis* and its name is put in shutter quotes (quotation marks) (Wiley, 1979).

One of the common problems you will encounter is the leftover taxa (candidates for *incertae sedis* under Convention 5) and "taxa" that are peripheral to your problem and are neither demonstrably monophyletic, paraphyletic, or (rarely) polyphyletic. Sometimes, these groups are well known. To omit them would leave the classification incomplete and make a statement that you know they are not monophyletic; to put them in without qualification would imply that you consider them monophyletic. The shutter quotes carry the connotation that all included subtaxa are presently *incertae sedis* at the place in the hierarchy where you put the "taxon." This convention should be used with caution. For example, it is not advisable to treat a polytomy occurring in the middle of a phylogenetic hypothesis (see Fig. 6.14) with this convention.

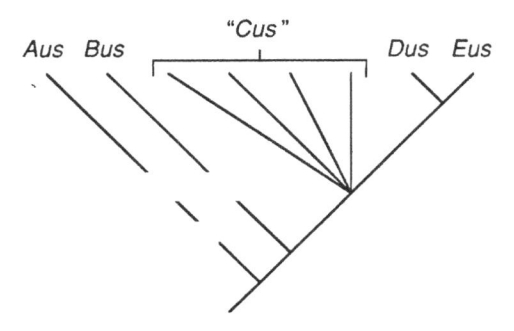

Fig. 6.14.—Relationships among a hypothetical group of genera, illustrating the unwise use of shutter quotes (Convention 6).

Convention 7.—Fossil taxa will be treated differently than Recent taxa. Fossil taxa will always be sequenced with their Recent relatives following Convention 3. If they are ranked, their status as fossils will be denoted by placing a dagger or cross symbol before the rank (Nelson, 1972). Alternatively, they may be given the neutral rank of "plesion" (Patterson and Rosen, 1977). As natural taxa, monophyletic fossil taxa may stand *incertae sedis* or *sedis mutabilis,* just as any Recent natural taxon.

Wiley's original conventions (Wiley, 1979, 1981a) incorporated only the plesion convention. The convention as stated here is less restrictive, but still keeps our ever-shifting understanding of the relationships of fossil groups from continually changing the hierarchical classifications of Recent groups. Figure 6.15 illustrates two ways of classifying some fossil mammals with their Recent relatives.

    a   Infradivision Theria
          Supercohort Marsupialia
          Supercohort Eutheria

b   Infradivision Theria   **c**   Infradivision Theria
    Plesion *Kueneotherium*   †Supercohort Kueneotheria
    Plesion Symmetrodonta   †Supercohort Symmetrodontia
    Plesion Dryolestoidea   †Supercohort Dryolestia
    Plesion *Paramus*   †Supercohort Paramia
    Supercohort Marsupialia   Supercohort Marsupialia
    Supercohort Eutheria   Supercohort Eutheria

Fig. 6.15.—Three classifications of some mammals (see Convention 7). a. Classification containing only groups with Recent species. b. "Major" fossil groups added to the Recent classification using the plesion convention. c. The same groups added but using the listing and dagger conventions.

Convention 8.—Stem species (ancestral species) are placed in classifications in parentheses beside the names of taxa they gave rise to or taxa containing their descendants, as appropriate. A stem species can be inserted into the hierarchy in one of three ways. 1) A stem species of a suprageneric taxon will be placed in a monotypic genus and inserted in the hierarchy beside the name of the taxon that contains its descendants. 2) A stem species of a genus will be placed in that genus and inserted beside the genus name. 3) The stem species of a species within a genus will be placed in that genus and inserted beside the species name.

Stem species, as Hennig (1966) recognized, are equivalent to the supraspecific taxa that contain their descendants. This convention treats stem species in exactly this manner. We illustrate this convention with the hypothetical ancestor, recently "discovered," of the Sarcopterygii (Fig. 6.16).

CLASSIFICATION

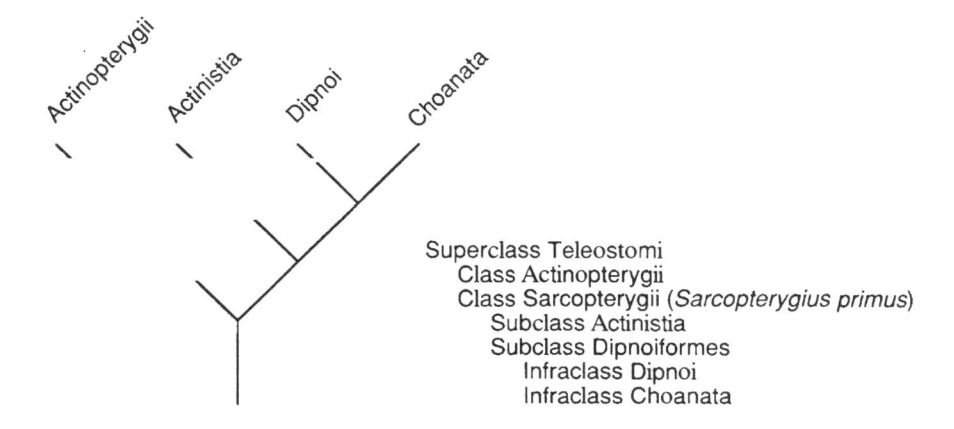

Fig. 6.16.—A phylogeny of teleostome vertebrates and a classification illustrating the ancestor convention (Convention 8).

Convention 9.—A taxon of hybrid origin will be indicated by placing the names of its parental species in parentheses beside the taxon's name. If the taxon is placed beside one of the parental taxa, its sequence in a list carries no connotation of branching sequence relative to taxa of "normal" origin.

Wiley (1981a:226–227) suggested that hybrids between species in different genera should be placed in a third genus. This is not necessary, but you should not place the hybrids in both genera if you do not wish to name a third genus because it might cause confusion regarding the actual number of species that exist. There are several ways this convention can be used. One way is shown in Fig. 6.17. You might also try listing the hybrid taxa under, say, the genus name and then proceed with the species of nonhybrid origin using other conventions.

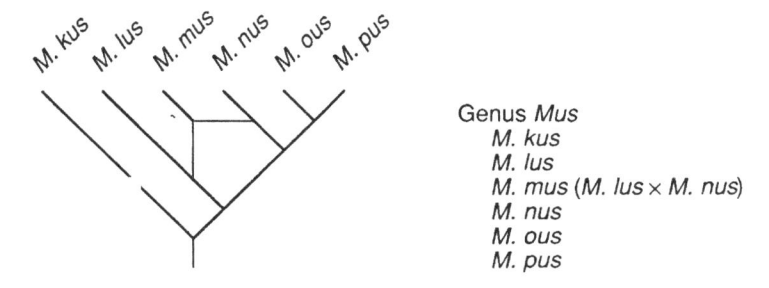

Fig. 6.17.—A phylogeny of the hypothetical genus *Mus*, with a classification illustrating the hybrid taxon convention (Convention 9).

The following classification of some species of the lizard genus *Cnemidophorus* shows a really nasty case handled in this manner.

Genus *Cnemidophorus*
  species of hybrid origin:
    C. uniparens (*inornatus* × *gularis* × *inornatus*)
    C. neomexicanus (*inornatus* × *tigris*)
    C. laredoensis (*sexlineatus* × *gularis*)
    C. "tesselatus" A–B ("*tesselatus*" *C–E* × *sexlineatus*)
    C. "tesselatus" C–E (*tigris* × *septemvittatus*)
  C. tigris species group
    C. tigris
  C. sexlineatus species group
    sexlineatus squadron
      C. sexlineatus
      C. inornatus
    gularis squadron
      C. gularis
      C. septemvittatus

This convention also works for taxa of symbiotic origin, except that the usual "×" sign that indicates hybrids is replaced by a "+" sign to indicate the "additive" nature of the resulting organism.

Quick Quiz—Taxonomy vs. Systematics

The Central Names Committee of the National Mugwort Society announces that it has accepted your recent phylogenetic study of spiny-toed mugworts, except that all of your higher order categories have been collapsed to subgenera. What should you do?

Convention Exercises

The following exercises are designed to give you some practice in using the conventions outlined above. These exercises are rather different from the exercises in the previous chapters because of the nature of the rules and the fact that ranks are relative. Also, note that the names of taxa are letters. You may use the letters as is, or if your instructor wishes, use them as the root to which a correct ending is added.

EXERCISE 6.3.—From the phylogenetic tree in Fig. 6.18, do the following:

1. Use the sequence convention at every opportunity to produce a classification beginning at the hierarchical rank of order. For higher *group* names, use the following convention.

CLASSIFICATION

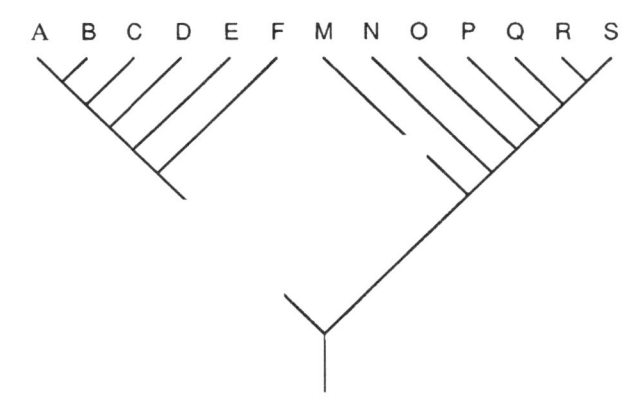

Fig. 6.18.—A phylogeny of the hypothetical order A–S (Exercise 6.3).

For a group composed of A, B, C, and D, use A–D; for a group composed of A and B, use AB, etc.

2. Produce a second classification without using the sequencing convention that is, by naming every branch point. For example, A–F is an order with two suborders, F and A–E. Don't forget to rank all sister groups at the same rank.

3. Count the number of hierarchical levels you save by using the sequencing convention.

EXERCISE 6.4.—From the phylogenetic tree of the hypothetical order A–U in Fig. 6.19, use Conventions 3, 4, and 5 to classify the members of the group using order, suborder, family, subfamily, and genus.

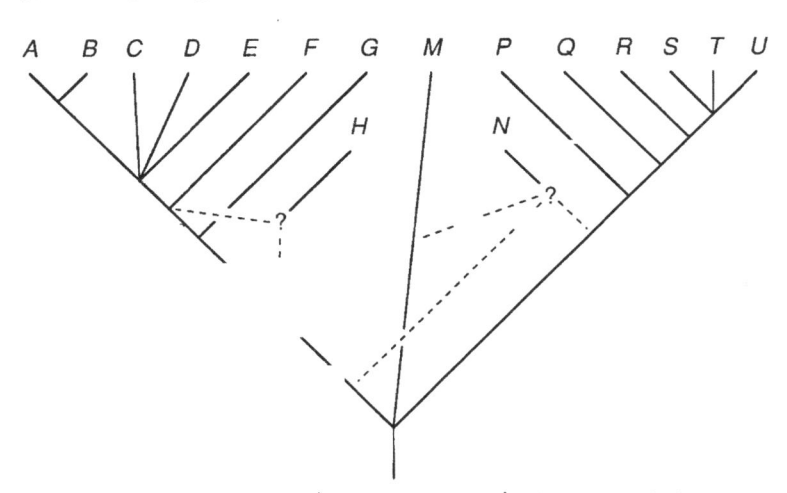

Fig. 6.19.—A phylogeny of the hypothetical order A–U (Exercise 6.4). The terminal taxa are genera.

EXERCISE 6.5.—From the phylogenetic tree in Fig. 6.20, do the following:

1. Naming every branch point, classify the fossil taxa **without** using Convention 7.
2. Classify the Recent taxa first and then add the fossil taxa into the classification using Convention 7
3. Briefly compare the two results in terms of hierarchical levels used and the magnitude of change in the classifications brought about by shifting the relationships of one fossil species.

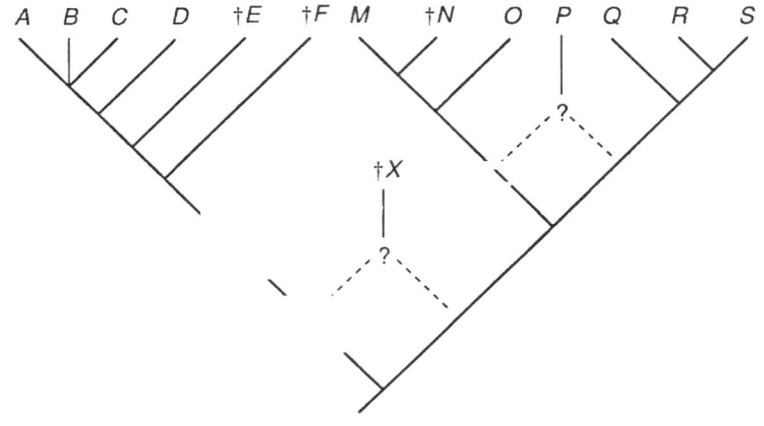

Fig. 6.20.—A phylogenetic hypothesis of the relationships among genera of the hypothetical order Tusiformes, including Recent and fossil groups (Exercise 6.5).

EXERCISE 6.6.—The very distinctive flightless cormorant.

The Galapagos Cormorant is very distinctive in that unlike the rest of the members of the family Phalacrocoracidae, it has small nonfunctional wings, peculiar wing feathers, and a massive bill. It was considered so distinctive that it was placed in its own genus, *Nannopterum* (literally "tiny wings").

1. Derive a phylogenetic tree from Table 6.2. The characters were polarized using the method of Madison et al. (1984) and the OG vector.
2. Evaluate the traditional classification below in terms of its consistency with the phylogeny.

Phalacrocorax africanus
  P. pygmaeus
  P. penicillatus
  P. carbo
  P. auritus
  P. olivaceus
Nannopterum harrisi

Table 6.2.—Data matrix for cormorant genera *Phalacrocorax* and *Nannopterum* (Exercise 6.6) (data from Siegel-Causey, 1988).

Taxon	Transformation series								
	8	15	17	30	51	80	82	86	93
OG	0	0		0	0	0	0	0	0
P. africanus	1	0		0	0	0	0	0	0
P. auritus	0	0		0	1	1	1	0	1
P. carbo	0	0		1	0	1	0	0	1
N. harrisi	0	1	0	0	0	1	0	1	0
P. olivaceus	0	0		0	1	1	0	0	1
P. penicillatus	0	0		0	0	1	0	1	0
P. pygmaeus	1	0	1	0	0	0	0	0	0

3. Construct a new classification if necessary.

CHAPTER NOTES AND REFERENCES

1. The criterion of logical consistency has also been discussed by Wiley (1981b, 1987a, 1989).

2. The literature on biological classification and its relationship to evolution is huge. The texts mentioned in Chapter 1, as well as the pages of *Systematic Zoology*, provide an introduction.

3. For an excellent introduction to some problems perceived in using the Linnaean system in phylogenetic classifications and a review of some of the proposed solutions, see Griffiths (1976). There are three basic options for adopting a set of conventions similar to that outlined here. First, one could simply use an unannotated Linnaean system of categories and name every branch. Second, one could adopt a numbering sequence (Hennig, 1981; Løvtrup, 1973). Third, one could abandon both numbering and prefixes and classify by pure indentation (cf. Gauthier et al., 1988).

QUICK QUIZ ANSWERS

Taxonomy vs. Systematics

Relax. It's the relationships that count for the systematist, not the names. Anyway, science by committee is for administrators and the faint hearted, not for you. Continue on as before.

Chapter 7

COEVOLUTIONARY STUDIES

Phylogenetic trees can be used in a variety of ways in evolutionary biology. Many of these applications involve comparing the degree to which the history of one group coincides with the history of the geographic areas in which its members reside or with the histories of other groups. Brooks (1985) suggested two basic reasons why a species might live where it lives or be associated with the particular species it is associated with. A species may live in a certain geographic area because its ancestor lived in that area and the descendant evolved there. Alternatively, the species may have evolved elsewhere and dispersed into the area where it now resides. In the first case (**association by descent**), we would expect the history of the species to coincide with the history of the area; whereas in the second (**association by colonization**), we would not. Likewise, two or more species that exist in a close ecological association may be associated because their ancestors were associated, or they may be associated because they evolved in association with other species and subsequently "switched allegiances." In the first case, we would expect the histories of the taxa involved in the association to coincide (to be **congruent**); whereas in the second case, we would not expect to find such congruence. Taxa that show historical congruence either with geographic areas or with other taxa are said to exhibit **cospeciation** patterns. Phylogenetic systematic methods can help distinguish associations due to ancestral association from associations due to dispersal and colonization.

In this chapter, we will learn how to use the results of phylogenetic analyses to study other aspects of the evolution of taxa. The major skills you will learn are 1) how to code the entire phylogenetic tree of a clade and 2) how to use the resulting matrix to study problems concerning biogeographic or coevolutionary aspects of the evolution of the clade or clades in the study.

Coding Phylogenetic Trees

Biogeographic and coevolutionary studies are concerned with the correlation of one set of data with another. For this correlation, the data can be partitioned into independent and dependent variables. For example, if we wish to study the relationships among a number of biogeographic areas based on the species that occur in these areas, we could consider the areas the independent variables and the phylogeny of the species the dependent variables. As another example, we might wish to assess the amount of cospeciation that has occurred during the evolution of a clade of parasites and their hosts. The relationships among the hosts would be considered the independent variables, and the relationships within the clade of parasites would be considered the dependent variables. The "fit" would be a function of the correlation between the host phylogeny (analyzed independently) and the parasite phylogeny. Another way we could use coded trees is to arrive at a hypothesis about the relationships of areas or hosts. We can code a tree in the following manner.

1. Label all nodes on the phylogenetic tree (Fig. 7.1). The tree will now consist of the branching structure, the terminal taxa, and labels for the ancestral nodes.

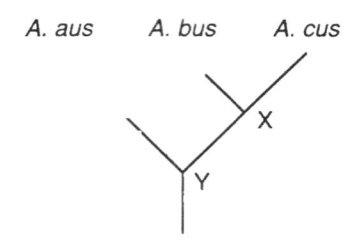

Fig. 7.1.—The relationships among three hypothetical species, with labels (X, Y) for their ancestral nodes.

2. Prepare a list of the occurrence of each terminal taxon relative to the independent variable. In this case the independent variables are areas of occurrence.

Area 1: *Aus aus*
Area 2: *Aus bus*
Area 3: *Aus cus*

3. Begin constructing a matrix (Table 7.1). By convention, rows are the independent variables (areas), and columns are the dependent variables (taxa). Each row is composed of binary values that can be inferred from the original phylogenetic tree. In our case, we have three rows (areas) and five columns (one for each branch and internode). Assign to a cell the value "1" if either a taxon or its ancestor can be inferred to occur or have occurred in the area. In biogeographic studies, the assumption is that the ranges of ancestors are the additive ranges of their descendants. The same assumptions are used in coevolutionary studies; for example, the range of an ancestral parasite is the additive range of its descendants relative to

Table 7.1.—Data matrix of the inferred distributions for taxa in Fig. 7.1.

Area	A. aus	A. bus	A. cus	X	Y
1	1	0	0	0	1
2	0	1	0	1	1
3	0	0	1	1	1

its hosts. In our simple example, Area 2 would be scored "1" for the *A. bus* and the X and Y₁ columns, reflecting the observed occurrence of *A. bus* in Area 2 and the inferred occurrence of X (*A. bus*'s ancestor) and Y₁ (the ancestor of the entire species group) in the same area. Let's take a closer look at this matrix. The first thing that should strike you is that there are three different sorts of data. 1) There are observational scores. Area 1 is scored "1" for *A. aus* as a matter of observation. 2) There are inferential scores. Area 1 has been scored "1" for hypothetical ancestor Y₁ because, under the assumption of additive ranges, we can infer that the ancestor of the clade was in Area 1 based on the presence of one of its descendants (*A. aus*). 3) There are negative observational scores. Area 1 is scored "0" for the *A. bus* cell because *A. bus* was not observed in Area 1. From this, we can infer that X was also absent from Area 1.

4. You now have a choice of methods, depending on what you know about the independent variables. If the relationships among the independent variables are not known, you can use the data to estimate these relationships. In this case, solve the tree for the dependent variable using the data from the independent variable. Do this just as if the coded tree were a matrix of characters for the areas. If an independent estimate of the independent variable is available, map the occurrence of taxa, including ancestors, on the tree that already exists for the areas. You can then use various tree statistics to judge the amount of coevolution or vicariance that can be inferred.

When you think about it, it doesn't make much sense to solve the relationships of the dependent variable with the information based on a single clade. Why? Because a single clade is to an independent variable what a single transformation series is to a clade. Therefore, using only single clades, we are restricted to an *a priori* hypothesis of the relationships among the independent variables. Before we discuss methods for handling more than one clade, let's look at a real example involving only one group.

Quick Quiz—Biogeography

1. What are the relationships of areas 1, 2, and 3 implied by the distributions of the three species of *Aus*?
2. If you only used "presence and absence" data, what would you conclude about the relationships of these areas?

Example 7.1.—Biogeography of the Amphilinidea.

Amphilinids are a small group of parasitic flatworms found on several continents (Bandoni and Brooks, 1987). Their phylogenetic relationships are shown in Fig. 7.2.

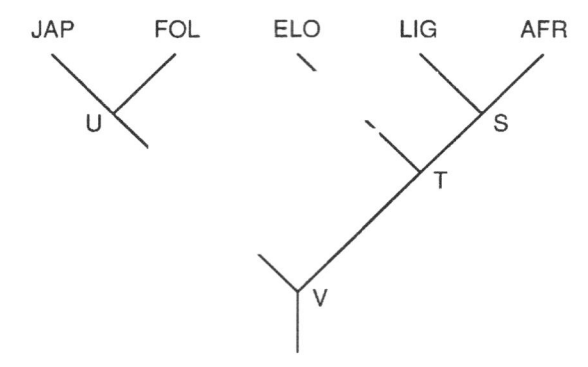

Fig. 7.2.—The relationships among five species of parasitic flatworms (from Bandoni and Brooks 1987) (Example 7.1). See text for abbreviations.

1. Label the nodes (already shown in Fig. 7.2).
2. Prepare a list showing the occurrence of each of the terminal taxa.

Amphilina japonica (JAP): North America (NA)
A. foliacea (FOL): Eurasia (EU)
Gigantolina elongata (ELO): Australia (AS)
Schizochoerus liguloideus (LIG): South America (SA)
S. africanus (AFR): Africa (AF)

3. Prepare a table of areas × taxa. Note that we have carefully rearranged the order of the taxa to put ancestral columns as close to relevant descendants as possible (Table 7.2). This is not necessary, but it helps when you enter data and proof the matrix.

4. We now have a choice. We can either solve the relationships of the areas or we can map the information in our matrix onto a previous hypothesis of area relationships. Or, we can do both and then compare the results. In this case, geologic evidence has been used to produce an "area cladogram" showing the historical connections among the areas (Fig. 7.3).

Table 7.2.—Data matrix of the inferred distributions of the Amphilinidea (Example 7.1).

Area*	JAP	FOL	U	LIG	AFR	S	ELO	T	V
NA	1	0	1	0	0	0	0	0	1
EU	0	1	1	0	0	0	0	0	1
SA	0	0	0	1	0	1	0	1	1
AF	0	0	0	0	1	1	0	1	1
AS	0	0	0	0	0	0	1	1	1

* NA = North America, EU = Eurasia, SA = South America, AF = Africa, AS = Australia.
† JAP = *Amphilina japonica*, FOL = *A. foliacea*, LIG = *Schizochoerus linguloideus*, AFR = *S. africanus*, ELO = *Gigantolina elongata*. S–V are ancestral taxa.

COEVOLUTIONARY STUDIES   117

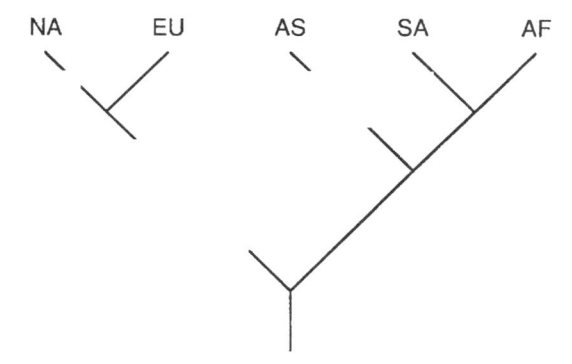

```
    NA     EU      AS     SA     AF
      \   / \       \      \    /
       \ /   \       \      \  /
        V     \       \      \/
               \       \     /
                \       \   /
                 \       \ /
                  \       V
                   \     /
                    \   /
                     \ /
                      V
                      |
```

Fig. 7.3.—The relationships among five continents based on geologic evidence and the relationships among amphilinid flatworms. See text for abbreviations.

When we solve the relationships among the areas based on the area/taxon matrix, we find the same hypothesis. Thus, the relationships among areas as determined from geologic data and from amphilinid phylogeny are congruent.

We may also estimate the level of congruence by mapping (fitting) the data for amphilinids directly onto the geologic hypothesis of the relationships among the continents. This approach treats the *tree* of continents as the independent variable. The matrix derived from the dependent variable (the continent × amphilinid matrix) could then be fitted to the hypothesis of continental relationships. This is easily accomplished in most computer programs such as PAUP, where we build a matrix, specify a particular tree (not necessarily derived from the matrix), and fit the matrix to the tree. We can then calculate an appropriate summary statistic. In this case, we select the CI, deleting the distributions of terminal taxa, and obtain a perfect fit (CI = 1.0).

Note on single group analysis. Although this kind of analysis was originally designed to investigate biogeographic and coevolutionary problems, you might find it suitable for exploring other kinds of problems. For example, you and Dr. Fenetico have reconstructed different phylogenies for the saber-toothed cnidaria. You might use your tree as the independent variable and Dr. Fenetico's data as the dependent variables. One very pleasing result might be the discovery that his data fit your tree better than they do his tree.

Single Tree Exercises

In the exercises below, 1) code the tree, and 2) solve the relationships among the independent variables. Do not be confused by distributions and area labels. Each species occupies a single area and can be given a single label regardless of the number of drainages and states in which it occurs. We are simply presenting real examples.

EXERCISE 7.1.—*Luxilus zonatus* species group.

This is a group of three species of North American minnows recently revised by Mayden (1988a).

Luxilus cardinalis: lower Arkansas River drainage, Arkansas, Kansas, and Oklahoma, and Red River tributaries of the Ouachita highlands, Oklahoma (Area 1)
L. pilsbryi: White and Little Red rivers, Missouri and Arkansas (Area 2)
L. zonatus: Ozark Plateau, north and east of *L. pilsbryi* in Missouri and Arkansas (Area 3)

Luxilus cardinalis is the sister of *L. pilsbryi*, and *L. zonatus* is the sister of this pair.

EXERCISE 7.2.—*Fundulus nottii* species group.

The *Fundulus nottii* species group consists of five North American topminnows with distributions shown below. (*Fundulus blairae* is actually more widespread than the information given, occurring along the northern Gulf Coast in sympatry with both *F. nottii* and eastern populations of *F. escambiae*.)

F. lineolatus (LIN): Florida peninsula and Atlantic Coast drainages (Area 1)
F. escambiae (ESC): eastern Gulf Coast (Area 2)
F. nottii (NOT): central Gulf Coast (Area 3)
F. blairae (BLA): lower Mississippi River and western Gulf Coast (Area 4)
F. dispar (DIS): upper Mississippi River (Area 5)

Wiley (1977) hypothesized the following relationships for these species:

((LIN(ESC, NOT))(BLA, DIS)).

More Than One Group

This method can also be used to compare the degree of congruence between geographic history and phylogeny for more than one group at a time. We treat each phylogenetic tree as a single chunk of the area × taxon matrix and perform a multigroup analysis. Analysis of multiple clades in the real world will almost certainly be more complicated than single clade analysis for two reasons. 1) Clades are not always found in all areas analyzed. For example, although family Aidae has members in all four of the areas you wish to study, family Bidae has members in only three of the areas. 2) Some clades may have members in two or more areas represented by single species in other clades. These are known as the "missing taxon" and "widespread species" problems. Both require us to consider character coding strategies that are not necessary when we perform single group analysis. Before attacking these problems, let us consider the simple example of two groups inhabiting the same region. We will use the classic hypothetical case presented by Humphries and Parenti (1986) and analyzed by Wiley (1987c, 1988a,b).

Example 7.2.—Two-group analysis.

Consider the hypothetical example below for two taxa, Lizard (L) and Frog (F), inhabiting the same areas (from Humphries and Parenti, 1986). The relationships of both groups are shown in Fig. 7.4. Their biogeographic ranges are listed below.

L1: Australia (AS)
L2: New Guinea (NG)
L3: South America (SA)
L4: Africa (AF)
F1: AS
F2: NG
F3: SA
F4: AF

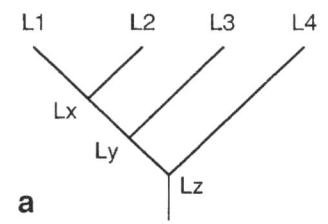

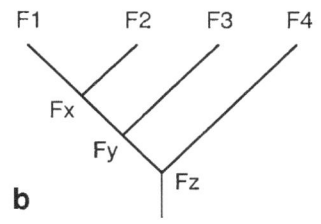

Fig. 7.4.—Relationships and labeled ancestral nodes for a group of (a) lizards (L1–L4) and a group of (b) frogs (F1–F4) (from Humphries and Parenti, 1986) (Example 7.2).

1. Label each ancestor on the tree of each group. (Fig. 7.4).
2. Produce a binary coded matrix just as you would for a single group, but with both groups in the matrix (Table 7.3).

Table 7.3.—Data matrix for lizards (L1–L4) and frogs (F1–F4) and their ancestral taxa (x, y, z) (Example 7.2).

Area*	L1	L2	Lx	L3	Ly	L4	Lz	F1	F2	Fx	F3	Fy	F4	Fz
AS	1	0	1	0	1	0	1	1	0	1	0	1	0	1
NG	0	1	1	0	1	0	1		1	1	0	1	0	1
SA	0	0	0	1	1	0	1		0	0	1	1	0	1
AF	0	0	0	0	0	1	1	0	0	0	0	0	1	1

* AS = Australia, NG = New Guinea, SA = South America, AF = Africa.

3. Solve the area relationships. To accomplish this, we must have an equivalent of character polarity decisions. This equivalent is the "zero vector ancestor" for each group. Specifically, we consider all "1" values as analogous to apomorphies by inspecting the outgroup and seeing that it does not occur in the area or by assuming that any distributional data we might gather that relate to events prior to the origin of the ancestor of each group is irrelevant. Just as in character analysis, the confidence we have in our hypothesis of area relationships depends on the confidence we have in our polarity assignments. The solution to this matrix is shown in Fig. 7.5. This is the history of vicariance based on phylogenetic and biogeographic data.

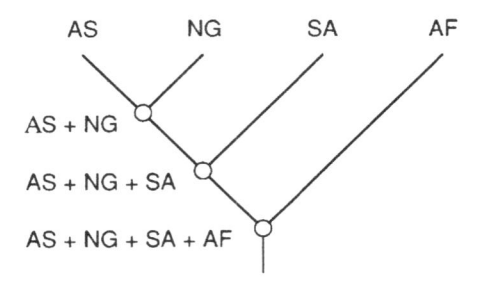

Fig. 7.5.—The relationships among four continents as shown by the lizard and frog phylogenetic and biogeographic data, with "ancestral areas" labeled at each node.

Missing Taxa

Not all clades are distributed over all areas. There are several reasons why a representative of a group may not be in an area. 1) The group might be absent because the ancestor of the group never inhabited the area. 2) The group might be absent because a member of the group went extinct in the area. 3) The group might really be present but was not sampled. The correct explanation for the absence of a group from an area will vary from group to group. Several strategies might be employed to deal with the problem.

Strategy 1.—Throw out the group. This strategy might be employed if one suspects that sampling effort is so low that the actual distributions of members of the group are not well enough known to employ them in an analysis.

Strategy 2.—Use the group. This strategy can be used if the investigator has reason to believe that the distribution data she has are robust enough to be employed. Sampling has been thorough enough so that the distribution of the group is fairly well to very well known. Any absence due to sampling error is probably randomized over all groups. Hypotheses of absence due to extinction or ancestral distribution (i.e., ancestor was not distributed over all areas included in the analysis) are viable explanations.

COEVOLUTIONARY STUDIES   121

Note the difference in the two strategies. If you employ the group but then conclude that the anomalies in its distribution relative to the general hypothesis you generate were due to sampling error, it would have been wiser not to use the group in the first place. Unfortunately, there are no hard and fast rules for employing either strategy. Strategy 2 is the best to use when there is any question about actual distributions. The only effect of introducing such a group, if the distribution of its members are at variance with the general hypothesis generated, should be an increase in tree length.

In Example 7.3, we will assume that you are satisfied that strategy 2 has been employed. Your assumption is that although absences may be due to sampling error, you have made some effort to discover members of groups that are missing in particular areas.

Example 7.3.—Missing taxa.

Taking the same areas as in Example 7.2, we now consider the distributional data provided by two additional groups, Bird (B) and Worm (W).

B1: Australia (AS)
B2: New Guinea (NG)
B4: Africa (AF)
W1: AS
W3: South America (SA)
W4: AF

Note that each group is missing a representative in one area. The relationships of these species are shown in Fig. 7.6. We proceed as follows.

1. Assign ancestors to the tree (Fig. 7.6).
2. Prepare the matrix (Table 7.4). The cells in the matrix that have dashes (–) are cells for which there are missing data. We have to decide how to code these missing data. Recall the earlier discussion about observational and inferential data scores. Turning to the Bx/SA cell in the matrix, if we decide to code the Bx/SA cell as "0," we are making the explicit inference that the ancestor of Bird was not present in South America. If we code it as "1," we

Fig. 7.6.—The relationships of a group of (a) birds (B) and (b) worms (W), with ancestral nodes labeled (x, y, z) (Example 7.3).

Table 7.4.—Data matrix for birds (B1, B2, B4) and worms (W1, W3, W4) and their ancestral taxa (x, y, z) (Example 7.3).

Area*	0					Taxa				0
	B1	B2	Bx	B4	Bz	W1	W3	Wy	W4	Wz
AS	1		1	0	1	1	0	1	0	1
NG	0	1	1	0	1	0	0	–	0	–
SA	0	0	–	0	–		1	1	0	1
AF	0	0	0	1	1	0	0	0	1	1

* AS = Australia, NG = New Guinea, SA = South America, AF = Africa.

are making the explicit inference that there is a member of Bird in South America or, alternatively, that there was a member of Bird there and it has gone extinct (i.e., we assign "1" as a phantom code indicating that we believe there is a missing ancestor and descendant). None of these alternatives are based on data nor on assumptions about the additive nature of descendant ranges relative to ancestral ranges. All are reasonable alternative hypotheses, at least at this stage in our analysis. We can remain neutral to these possibilities by assigning a missing data code ("?") to this cell. We proceed to each cell in the matrix and assign the missing data code to each ancestral column when a member of that group is not present in the area (Table 7.5).

3. We then solve for the area relationships. When doing this we must take into account the "missing data" cells and their alternative interpretations under different kinds of optimization. Figure 7.7 shows three equally parsimonious interpretations using DELTRAN. Figure 7.7a shows the effect of delaying the "appearance" of Bx. If this is our choice, then we cannot delay the appearance of Wy. Figure 7.7b shows the effect of delaying the "appearance" of Wy. If we delay Wy, we cannot delay Bx. Finally, Fig. 7.7c summarizes the conflict between the distributions of both characters. Note that in neither case is there **direct** evidence for the grouping AS+NG+SA. Rather, the grouping is a by-product of the interactions between character distributions. Also note that even using DELTRAN neither Bz nor Wz is effectively delayed. This is another interaction effect. (If you try ACCTRAN, you will find that both Bx and Wy are successfully accelerated but that the effect is to provide no character support for the groupings AS+NG or AS+SA.)

Table 7.5.—Data matrix from Table 7.4 modified to show neutrality about missing data.

Area	0					Taxa			0	
	B1	B2	Bx	B4	Bz	W1	W3	Wy	W4	Wz
AS	1	0	1		1	1	0	1		1
NG	0	1	1	0	1	0	0	?	0	?
SA	0	0	?	0	?	0	1	1	0	1
AF	0	0	0	1	1	0	0	0	1	1

COEVOLUTIONARY STUDIES   123

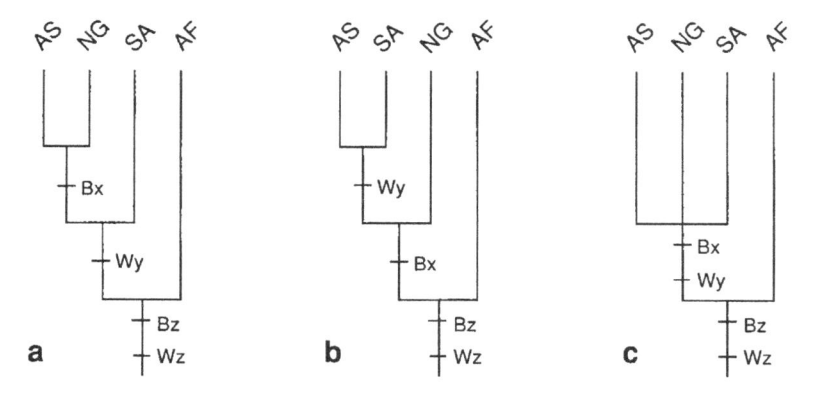

Fig. 7.7.—Three alternative trees (a–c) derived from analyzing the bird and worm data. Note that the synapomorphies supporting AS+NG (tree a) and AS+SA (tree b) are illusions resulting from "missing data" cells in the matrix.

EXERCISE 7.3.—Some moths.

Examine the distribution data listed below and the phylogeny shown in Fig. 7.8.

1. Prepare a data matrix for this group using the original four areas (AS, NG, SA, AF).

M1: AS
M2: NG
M3: SA

2. Add these data to the data concerning Bird and Worm (Table 7.5) and solve the area relationship problem.

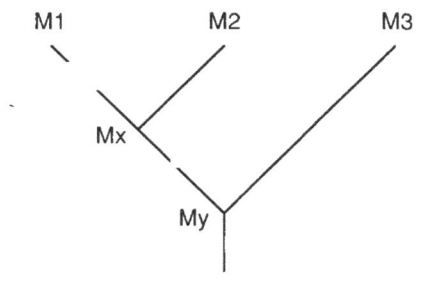

Fig. 7.8.—The relationships among three moths (M) with ancestral nodes labeled (x, y) (Exercise 7.3).

Widespread Species

A widespread species is one that is found in two or more areas. Obviously, this designation is relative to the distributions of species in other groups. A group with a widespread species shows less endemism than a group with different species in all the areas analyzed. A species might be widespread for several reasons. 1) The species did not respond to geographic subdivision by speciating (failure to speciate). 2) The species dispersed into one of the areas (dispersal). 3) The species is really two species, but the investigator has not detected this fact (identification error).

Given that you have analyzed the phylogenetic relationships among members of the group carefully and have done your best to identify each species, the only coding strategy you can employ is to code each species where it occurs. Example 7.4 demonstrates how to code for such cases.

Example 7.4.—Coding widespread taxa.

Continuing our example from Humphries and Parenti (1986), we consider two groups having widespread species: Tree (T) and Fish (Fi). The phylogenies of the two groups are shown in Fig. 7.9. The distribution of species in each group is listed below.

T1: Australia (AS)
T2: New Guinea (NG)
T34: South America (SA), Africa (AF)
Fi14: AS, AF
Fi2: NG
Fi3: SA

We proceed as follows.

1. Assign ancestors to the trees (Fig. 7.9).
2. Prepare the data matrix (Table 7.6). In this case, we will not assume that widespread species are either useful or not useful for solving the area relationships. Therefore, both the area and the ancestor will be assigned "1" when a descendant is present.

Fig. 7.9.—The relationships within a group of (a) trees (T) and (b) fishes (Fi), with ancestral nodes labeled (x, y) (Example 7.4).

Table 7.6.—Data matrix for trees (T1, T2, T34) and fish (Fi14, Fi2, Fi3) and their ancestral taxa (x, y) (Example 7.4).

Area*	Taxa								
	T1	T2	Tx	T34	Ty	Fi14	Fi2	Fix	Fi3	Fiy
AS	1	0	1	0	1	1	0	1	0	1
NG	0	1	1	0	1	0	1	1	0	1
SA	0	0	0	1	1	0	0	0	1	1
AF	0	0	0	1	1	1	0	1	0	1

* AS = Australia, NG = New Guinea, SA = South America, AF = Africa.

3. Solve the area relationships. In this case, there are three possible trees, each with 15 steps and a CI of 0.833. Rather than present one or more of these trees, we are going to map the distributions of members of Tree and Fish onto the area relationships shown by Lizard and Frog (Example 7.2). This result is shown in Fig. 7.10. (Considering only the Tree and Fish data, this tree is 13 steps and has a CI of 0.769.)

The tree in Fig. 7.10 reveals some interesting problems. Let us consider the distributions of T34, Fix, and Fi14 in the light of the three reasons for widespread species listed above. Reasons 1 and 3 lead you to expect that the species can be mapped as monophyletic on the area cladogram. In neither case, however, will the outcome affect the area cladogram. Reasons 1 and 2 might be expected if the species in question is paraphyletic when mapped on the area cladogram. Reason 1 is a little scary because acceptance of the hypothesis leads to accepting the species as an ancestor. Reason 2 would be expected if the species is polyphyletic when mapped on the area cladogram. We can be fairly confident that Fish 14 has dispersed from Australia to Africa, but we are not very confident that Tree 34 dispersed or simply failed to respond to the vicariance event that separated South America and Africa.

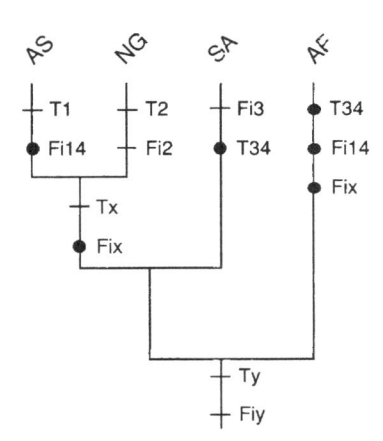

Fig. 7.10.—A hypothesis of the relationships among four areas with the occurrence of trees and fishes mapped on the hypothesis. Tick marks denote unique distributions, black dots denote homoplasy.

EXERCISE 7.4.—Ferns.

Examine the distributional data and phylogeny (Fig. 7.11) for the group Fern (Fe). Prepare a data matrix for this group. Combine this with the data matrix for Tree and Fish and solve the area relationships using DELTRAN. **Tip:** There are three DELTRAN trees and three ACCTRAN trees, but solve only for DELTRAN.

Fe12: Australia (AS), New Guinea (NG)
Fe3: South America (SA)
Fe4: Africa (AF)

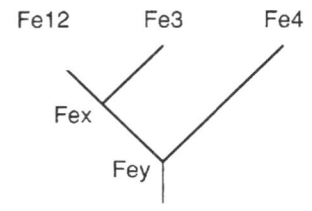

Fig. 7.11.—The relationships among three species of ferns (Fe), with ancestral nodes labeled (x y) (Exercise 7.4).

EXERCISE 7.5.—Combining the matrix.

Combine the matrices for all of the groups inhabiting the four continental regions we have been working with (a total of eight groups). Solve the area relationships using DELTRAN. Examine the distribution of each group on the independent variable (the cladogram of areas) and see if you can detect cases where a strict vicariance interpretation should not be followed.

Sympatry within a Clade

Sympatry within a clade occurs when two members of the clade inhabit the same area. How to treat sympatry between members of the same clade is still an open issue, and we have not designed exercises for this aspect of vicariance biogeography. Three suggestions have been made. Wiley (1988a) simply coded sympatric members of a clade as being present in the area. Kluge (1988b) suggested that in such cases one of the distributional patterns is younger than the origin of the areas. Thus, one of the members of the clade cannot furnish corroboration for the area hypothesis. This is certainly true, but the problem is, which member? Brooks (1990) has suggested another strategy, assigning two codes to the area where two members of the clade are sympatric. So, if *Aus aus* and *Aus bus* both inhabit Area

1, split Area 1 into two areas (1 and 1'). Then find out where 1' falls out in the analysis. How the two species are related to each other and how the two areas are related to each other can then be addressed.

THE ANALOGY BETWEEN PHYLOGENETICS AND HISTORICAL BIOGEOGRAPHY

You can see that the computational similarities among phylogenetic systematics, vicariance biogeography, and studies of coevolution can be very impressive. However, the biological basis behind each area of study is different. For example, the relationship between characters and taxa is much more direct than the relationship between areas and taxa. Look at Fig. 7.10 again. Note that one of the ancestors in the Fish group is actually plotted as polyphyletic (!). In character analysis we would simply conclude that the distribution was achieved by convergent evolution. This is allowed. However, the origin of a single species twice is not allowed. Several authors have called attention to these problems, including Wiley (1987c), Cracraft (1988), Sober (1988b), and Brooks (1990).

CHAPTER NOTES AND REFERENCES

1. The use of a matrix to describe the shape of a tree was independently proposed by Farris (1969), Phipps (1971), and Williams and Clifford (1971); see Farris (1973).

2. We have used additive binary coding throughout this chapter. This is not necessary. Indeed, it is not necessary to enter the terminal taxa into a matrix because these data do not affect tree topology (cf. Kluge, 1988b; Mayden, 1988b). Further, we suggest that when you use a suitable computer program, you experiment with inputting your transformation series in such a manner that evolution is irreversible. See Mayden (1988b) for a discussion of possible problems in coding characters.

3. Brooks (1981) first proposed a technical solution to Hennig's parasitological method, and this led directly to the vicariance biogeographic and coevolutionary techniques presented here.

4. Biogeographic and coevolutionary techniques are virtually identical, as shown by Brooks (1985, 1988) and Wiley (1987d). A considerable body of literature on vicariance biogeographic methods and applications is summarized in Wiley (1988a). Parsimony methods are discussed in detail by Wiley (1987c, 1988a,b), Zandee and Roos (1987), and Kluge (1988b). Application of vicariance methods to the study of speciation include Wiley (1980, 1981a), Cracraft (1983, 1986), Mayden (1985, 1988b), Wiley and Mayden (1985), Brooks and McLennan (1991), and Siegel-Causey (1991).

5. See McLennan et al. (1988) for an example of the use of behavioral characters. For a review of a more general use of these methods in coevolutionary studies, see Brooks and McLennan (1991).

Quick Quiz Answers

Biogeography

1. (2+3)+(1). This is a short way to note the relationships.

2. Presence/absence data use only the first three columns of the data matrix (Table 7.1). When you use only these data, you see that the area relationships are not resolved. So the tree would be 1+2+3. Not very helpful, is it?

LITERATURE CITED

Adams, E. N. 1972. Consensus techniques and the comparison of taxonomic trees. Syst. Zool. 21:390–397.

Archie, J. W. 1989. Homoplasy excess ratios: New indices for measuring levels of homoplasy in phylogenetic systematics and a critique of the consistency index. Syst. Zool. 38:253–269.

Archie, J. W. 1990. Homoplasy excess statistics and retention indices: A reply to Farris. Syst. Zool. 39:169–174.

Ax, P. 1987. The Phylogenetic System. The Systematization of Organisms on the Basis of Their Phylogenesis. John Wiley & Sons, New York.

Bandoni, S. M., and D. R. Brooks. 1987. Revision and phylogenetic analysis of the Amphilinidae Poche, 1922 (Platyhelminthes: Cercomeria: Cercomeromorpha). Can. J. Zool. 65:1110–1128.

Bremer, K. 1978. The genus *Leysera*. (Compositae). Bot. Not. 131:369–383.

Bremer, K. 1985. Summary of green plant phylogeny and classification. Cladistics 1:369–385.

Brooks, D. R. 1981. Hennig's parasitological method: A proposed solution. Syst. Zool. 30:229–249.

Brooks, D. R. 1985. Historical ecology: A new approach to studying the evolution of ecological associations. Ann. Mo. Bot. Gard. 72:660–680.

Brooks, D. R. 1988. Macroevolutionary comparisons of host and parasite phylogenies. Ann. Rev. Ecol. Syst. 19:235–259.

Brooks, D. R. 1990. Parsimony analysis in historical biogeography and coevolution: Methodological and theoretical update. Syst. Zool. 39:14–30.

Brooks, D. R., J. N. Caira, T. R. Platt, and M. H. Pritchard. 1984. Principles and methods of cladistic analysis: A workbook. Univ. Kans. Mus. Nat. Hist. Spec. Publ. 12:1–92.

Brooks, D. R., and D. A. McLennan. 1991. Phylogeny, Ecology, and Behavior: A Research Program in Comparative Biology. University of Chicago Press, Chicago.

Brooks, D. R., R. T. O'Grady, and E. O. Wiley. 1986. A measure of the information content of phylogenetic trees, and its use as an optimality criterion. Syst. Zool. 35:571–581.

Brooks, D. R., and E. O. Wiley. 1985. Theories and methods in different approaches to phylogenetic systematics. Cladistics 1:1–11.

Brooks, D. R., and E. O. Wiley. 1988. Evolution as Entropy. Towards a Unified Theory of Biology. University of Chicago Press, Chicago.

Brundin, L. 1966. Transantarctic relationships and their significance, as evidenced by chironomid midges. Kungl. Svenska Vetenskap. Hamdl. 11:1–472.

Buth, D. G. 1984. The application of electrophoretic data in systematic studies. Annu. Rev. Ecol. Syst. 15:501–522.

Carpenter, J. M. 1988. Choosing among multiple equally parsimonious cladograms. Cladistics 4: 291–296.

Cracraft, J. 1983. Species concepts and speciation analysis. Curr. Ornithol. 1:159–187.

Cracraft, J. 1986. Origin and evolution of continental biotas: Speciation and historical congruence within the Australian avifauna. Evolution 40:977–996.

Cracraft, J. 1988. Deep-history biogeography: Retrieving the historical pattern of evolving continental biotas. Syst. Zool. 37:221–236.

Crisci, J. V., and T. F. Stuessy. 1980. Determining primitive character states for phylogenetic reconstruction. Syst. Bot. 5:112–135.

Cronquist, A. 1987. A botanical critique of cladism. Bot. Rev. 53:1–52.

Crowson, R. A. 1970. Classification and Biology. Heineman Education Books, London.

DeBry, R. W., and N. A. Slade. 1985. Cladistic analysis of restriction endonuclease cleavage maps within a maximum-likelihood framework. Syst. Zool. 34:21–34.

de Jong, R. 1980. Some tools for evolutionary and phylogenetic studies. Z. Zool. Syst. Evolutionsforsch. 18:1–23.

Donoghue, M. J., and P. D. Cantino. 1984. The logic and limitations of the outgroup substitution method for phylogenetic reconstructions. Syst. Bot. 9:112–135.

Dupuis, C. 1984. Willi Hennig's impact on taxonomic thought. Annu. Rev. Ecol. Syst. 15:1–24.

Eldredge, N., and J. Cracraft. 1980. Phylogenetic Patterns and the Evolutionary Process. Columbia University Press, New York.

Farris, J. S. 1969. A successive weighting approximations approach to character analysis. Syst. Zool. 18:374–385.

Farris, J. S. 1970. Methods of computing Wagner trees. Syst. Zool. 19:83–92.

Farris, J. S. 1972. Estimating phylogenetic trees from distance matrices. Am. Nat. 106:645–668.

Farris, J. S. 1973. On comparing the shapes of taxonomic trees. Syst. Zool. 22:50–54.

Farris, J. S. 1977. Phylogenetic analysis under Dollo's Law. Syst. Zool. 26:77–88.

Farris, J. S. 1979. The information content of the phylogenetic system. Syst. Zool. 28:483–519.

Farris, J. S. 1981. Distance data in phylogenetic analysis, pp. 2–23. In: Advances in Cladistics (V. A. Funk and D. R. Brooks, eds.). New York Botanical Garden, New York.

Farris, J. S. 1982. Outgroups and parsimony. Syst. Zool. 31:328–334.

Farris, J. S. 1983. The logical basis of phylogenetic analysis, pp. 7–36. In: Advances in Cladistics, 2 (N. I. Platnick and V. A. Funk, eds.). Columbia University Press, New York.

Farris, J. S. 1985. Distance data revisited. Cladistics 1:67–85.

Farris, J. S. 1986. Distances and statistics. Cladistics 2:144–157.

Farris, J. S. 1989a. Hennig86. Port Jefferson Station, New York.

LITERATURE CITED

Farris, J. S. 1989b. The retention index and rescaled consistency index. Cladistics 5:417–419.

Farris, J. S. 1990. The retention index and homoplasy excess. Syst. Zool. 38:406–407.

Farris, J. S., A. G. Kluge, and M. J. Eckhart. 1970. A numerical approach to phylogenetic systematics. Syst. Zool. 19:172–189.

Felsenstein, J. 1973a. Maximum likelihood estimation of evolutionary trees from continuous characters. Am. J. Hum. Genet. 25:471–492.

Felsenstein, J. 1973b. Maximum likelihood and minimum-step methods for estimating evolutionary trees from data on discrete characters. Syst. Zool. 22:240–249.

Felsenstein, J. 1977. The number of evolutionary trees. Syst. Zool. 27:27–33.

Felsenstein, J. 1978. Cases in which parsimony or compatibility will be positively misleading. Syst. Zool. 27:401–410.

Felsenstein, J. 1981. Evolutionary trees from gene frequencies and quantitative characters: Finding maximum likelihood estimates and testing hypotheses. Evolution 35:1229–1246.

Felsenstein, J. 1983. Parsimony in systematics: Biological and statistical issues. Annu. Rev. Ecol. Syst. 14:313–333.

Felsenstein, J. 1984. Distance methods for inferring phylogenies: A justification. Evolution 38:16–24.

Felsenstein, J. 1985. Confidence limits on phylogenies: An approach using bootstrap. Evolution 39:783–791.

Felsenstein, J. 1986. Distance methods: A reply to Farris. Cladistics 2:130–143.

Fitch, W. M. 1971. Toward defining the course of evolution: Minimum change for a specific tree topology. Syst. Zool. 20:406–416.

Funk, V. A. 1985 Phylogenetic patterns and hybridization. Ann. Mo. Bot. Gard. 72:681–715.

Gauthier, J., R. Estes, and K. de Queiroz. 1988. A phylogenetic analysis of Lepidosauromorpha, pp. 15–98. In: Phylogenetic Relationships of the Lizard Families. Essays Commemorating Charles L. Camp (R. Estes and G. Pregill, eds.). Stanford University Press, Stanford, California.

Griffiths, G. C. D. 1976. The future of Linnaean nomenclature. Syst. Zool. 25:168–173.

Hendy, M. D., and D. Penny. 1982. Branch and bound algorithms to determine minimal evolutionary trees. Math. Biosci. 59:277–290.

Hennig, W. 1950. Grundzüge einer Theorie der Phylogenetischen Systematik. Deutscher Zentralverlag, Berlin.

Hennig, W. 1953. Kritische Bermerkungen zum phylogenetischen System der Insekten. Beitr. Entomol. 3:1–85.

Hennig, W. 1965. Phylogenetic systematics. Annu. Rev. Entomol. 10:97–116.

Hennig, W. 1966. Phylogenetic Systematics. University of Illinois Press, Urbana.

Hennig, W. 1981. Insect Phylogeny. John Wiley & Sons, New York.

Hillis, D. M. 1984. Misuse and modification of Nei's genetic distance. Syst. Zool. 33:238–240.

Hillis, D. M. 1985. Evolutionary genetics of the Andean lizard genus *Pholidobolus* (Sauria: Gymnophthalmidae): Phylogeny, biogeography, and a comparison of tree construction techniques. Syst. Zool. 34:109–126.

Hillis, D. M., and C. Moritz. 1990. Molecular Systematics. Sinauer, Sunderland, Massachusetts.

Hull, D. L. 1964. Consistency and monophyly. Syst. Zool. 13:1–11.

Humphries, C. J., and L. Parenti. 1986. Cladistic biogeography. Oxford Monogr. Biogeogr. 2:1–98.

Kluge, A. G. 1984. The relevance of parsimony to phylogenetic inference, pp. 24–38. In: Cladistics: Perspectives on the Reconstruction of Evolutionary History (T. Duncan and T. Steussey, eds.). Columbia University Press, New York.

Kluge, A. G. 1985. Ontogeny and phylogenetic systematics. Cladistics 1:13–27.

Kluge, A. G. 1988a. The characterization of ontogeny, pp. 57–81. In: Ontogeny and Systematics (C. J. Humphries, ed.). Columbia University Press, New York.

Kluge, A. G. 1988b. Parsimony in vicariance biogeography: A quantitative method and a Greater Antillean example. Syst. Zool. 37:315–328.

Kluge, A. J. 1989. A concern for evidence and a phylogenetic hypothesis of relationships among *Epicrates* (Boidae, Serpentes). Syst. Zool. 38:7–25.

Kluge, A. G., and J. S. Farris. 1969. Quantitative phyletics and the evolution of anurans. Syst. Zool. 18:1–32.

Kluge, A. G., and R. E. Strauss. 1986. Ontogeny and systematics. Annu. Rev. Ecol. Syst. 16:247–268.

Løvtrup, S. 1973. Epigenetics. A Treatise on Theoretical Biology. John Wiley & Sons, New York.

Mabee, P. M. 1989. An empirical rejection of the ontogenetic polarity criterion. Cladistics 5:409–416.

Maddison, W. P., M. J. Donoghue, and D. R. Maddison. 1984. Outgroup analysis and parsimony. Syst. Zool. 33:83–103.

Maddison, W. P., and D. R. Maddison. In press. MacClade. Sinauer, Sunderland, Massachusetts.

Margush, T., and F. R. McMorris. 1981. Consensus n-trees. Bull. Math. Biol. 43:239–244.

Mayden, R. L. 1985. Biogeography of Ouachita Highlands fishes. Southwest. Nat. 30:195–211

Mayden, R. L. 1988a. Systematics of the *Notropis zonatus* species group, with description of a new species from the interior highlands of North America. Copeia 1984:153–173.

Mayden, R. L. 1988b. Vicariance biogeography, parsimony, and evolution in North American freshwater fishes. Syst. Zool. 37:329–355.

McLennan, D. A., D. R. Brooks, and J. D. McPhail. 1988. The benefits of communication between comparative ethology and phylogenetic systematics: A case study using gasterosteid fishes. Can. J. Zool. 66:2177–2190.

Mickevich, M. F. 1982. Transformation series analysis. Syst. Zool. 31:461–478.

Nelson, G. J. 1972. Phylogenetic relationship and classification. Syst. Zool. 21:227–231.

Nelson, G. J. 1978. Ontogeny, phylogeny, paleontology, and the biogenetic law. Syst. Zool. 27:324–345.

Nelson, G. 1979. Cladistic analysis and synthesis: Principles and definitions, with a historical note on Adanson's Familles des Plantes (1763–1764). Syst. Zool. 28:1–21.

Nelson, G. J. 1985. Outgroups and ontogeny. Cladistics 1:29–45.

Nelson, G. J., and N. I. Platnick. 1981. Systematics and Biogeography: Cladistics and Vicariance. Columbia University Press, New York.

O'Grady, R. T. 1985. Ontogenetic sequences and the phylogenetics of parasitic flatworm life cycles. Cladistics 1:159–170.

O'Grady, R. T., and G. B. Deets. 1987. Coding multistate characters, with special reference to the use of parasites as characters of their hosts. Syst. Zool. 36:268–279.

O'Grady, R. T., G. B. Deets, and G. W. Benz. 1989. Additional observations on nonredundant linear coding of multistate characters. Syst. Zool. 38:54–57.

Page, R. D. M. 1987. Graphs and generalized tracks: Quantifying Croizat's panbiogeography. Syst. Zool. 36:1–17.

Page, R. D. M. 1988. Quantitative cladistic biogeography: Constructing and comparing area cladograms. Syst. Zool. 37:254–270.

Page, R. D. M. 1989. Comments on component-compatibility in historical biogeography. Cladistics 5:167–182.

Parenti, L. R. 1981. A phylogenetic and biogeographic analysis of cyprinodontiform fishes (Teleostei: Atherinomorpha). Bull. Am. Mus. Nat. Hist. 168:335–557.

Patterson, C. 1982. Morphological characters and homology, pp. 21–74. In: Problems of Phylogenetic Reconstruction (K. A. Joysey, ed.). Academic Press, London.

Patterson, C., and D. E. Rosen. 1977. Review of ichthyodectiform and other Mesozoic fishes and the theory and practice of classifying fossils. Bull. Am. Mus. Nat. Hist. 158:81–172.

Phipps, J. B. 1971. Dendrogram topology. Syst. Zool. 20:306–308.

Pimentel, R. A., and R. Riggins. 1987. The nature of cladistic data. Cladistics 3:201–209.

Popper, K. R. 1965. The Logic of Scientific Discovery. Harper Torchbooks, New York.

Ridley, M. 1985. Evolution and Classification. The Reformation of Cladism. Long, New York.

Rieppel, O. 1985. Ontogeny and the hierarchy of types. Cladistics 1:234–246.

Sanderson, M. J. 1989. Confidence limits on phylogenies: The bootstrap revisited. Cladistics 5:113–129.

Sanderson, M. J., and M. J. Donoghue. 1989. Patterns of variation in levels of homoplasy. Evolution 43:1781–1795.

Schoch, R. M. 1986. Phylogeny Reconstruction in Paleontology. Van Nostrand Reinhold, New York.

Siegel-Causey, D. 1988. Phylogeny of the Phalacrocoracidae. Condor 90:885–905.

Siegel-Causey, D. 1991. Systematics and biogeography of North Pacific shags, with a description of a new species. Univ. Kans. Mus. Nat. Hist. Occ. Pap. 140:1–17.

Sober, E. 1983. Parsimony methods in systematics, pp. 37–47. In: Advances in Cladistics, vol. 2 (N. I. Platnick and V. A. Funk, eds.). Columbia University Press, New York.

Sober, E. 1988a. Reconstructing the Past: Parsimony, Evolution, and Inference. MIT Press, Cambridge, Massachusetts.

Sober, E. 1988b. The conceptual relationship of cladistic phylogenetics and vicariance biogeography. Syst. Zool. 37:245–253.

Sokal, R. R., and F. J. Rohlf. 1981. Taxonomic congruence in the Leptopodomorpha reexamined. Syst. Zool. 30:309–325.

Stevens, P. F. 1980. Evolutionary polarity of character states. Annu. Rev. Ecol. Syst. 11:333–358.

Swofford, D. L. 1990. Phylogenetic Analysis Using Parsimony (PAUP), version 3.0. Illinois Natural History Survey, Urbana.

Swofford, D. L., and S. H. Berlocher. 1987. Inferring evolutionary trees from gene frequency data under the principle of maximum parsimony. Syst. Zool. 36:293–325.

Swofford, D. L., and W. P. Maddison. 1987. Reconstructing ancestral character states under Wagner parsimony. Math. Biosci. 87:199–229.

Swofford, D. L. and G. J. Olsen. 1990. Phylogeny reconstruction, pp. 411–501. In: Molecular Systematics (D. M. Hillis and C. Moritz, eds.). Sinauer, Sunderland, Massachusetts.

Thompson, E. A. 1986. Likelihood and parsimony: Comparison of criteria and solutions. Cladistics 2:43–52.

Watrous, L. E., and Q. D. Wheeler. 1981. The out-group comparison method of character analysis. Syst. Zool. 30:1–11.

Weston, P. H. 1988. Indirect and direct methods in systematics, pp. 27–56. In: Ontogeny and Systematics (C. J. Humphries, ed.). Columbia University Press, New York.

Wiley, E. O. 1975. Karl R. Popper, systematics, and classification—A reply to Walter Bock and other evolutionary taxonomists. Syst. Zool. 24:233–243.

Wiley, E. O. 1977. The phylogeny and systematics of the *Fundulus nottii* group (Teleostei: Cyprinodontidae). Univ. Kans. Mus. Nat. Hist. Occ. Pap. 66:1–31

Wiley, E. O. 1979. An annotated Linnaean hierarchy, with comments on natural taxa and competing systems. Syst. Zool. 28:308–337.

Wiley, E. O. 1980. Parsimony analysis and vicariance biogeography. Syst. Bot. 37:271–290.

Wiley, E. O. 1981a. Phylogenetics. The Principles and Practice of Phylogenetic Systematics. John Wiley & Sons, New York.

Wiley, E. O. 1981b. Convex groups and consistent classifications. Syst. Bot. 6:346–358.

Wiley, E. O. 1987a. The evolutionary basis for phylogenetic classification, pp. 55–64. In: Systematics and Evolution: A Matter of Diversity (P. Hovenkamp, E. Gittenberger, E. Hennipman, R. de Jong, M. C. Roos, R. Sluys, and M. Zandee, eds.). Utrecht University, Utrecht, The Netherlands.

Wiley, E. O. 1987b. Approaches to outgroup comparison, pp. 173–191. In: Systematics and Evolution: A Matter of Diversity (P. Hovenkamp, E. Gittenberger, E. Hennipman, R. de Jong, M. C. Roos, R. Sluys, and M. Zandee, eds.). Utrecht University, Utrecht, The Netherlands.

Wiley, E. O. 1987c. Methods in vicariance biogeography, pp. 283–306. In: Systematics and Evolution: A Matter of Diversity (P. Hovenkamp, E. Gittenberger, E. Hennipman, R. de Jong, M. C. Roos, R. Sluys, and M. Zandee, eds.). Utrecht University, Utrecht, The Netherlands.

Wiley, E. O. 1987d. Historical ecology and coevolution, pp. 331–341. In: Systematics and Evolution: A Matter of Diversity (P. Hovenkamp, E. Gittenberger, E. Hennipman, R. de Jong, M. C. Roos, R. Sluys, and M. Zandee, eds.). Utrecht University, Utrecht, The Netherlands.

Wiley, E. O. 1988a. Vicariance biogeography. Annu. Rev. Ecol. Syst. 19:513–542.

Wiley, E. O. 1988b. Parsimony analysis and vicariance biogeography. Syst. Zool. 37:271–290.

Wiley, E. O. 1989. Kinds, individuals, and theories, pp. 289–300. In: What the Philosophy of Biology Is (M. Reese, ed.). Kluwer Academic Publ., Dordrecht, The Netherlands.

Wiley, E. O., and R. L. Mayden. 1985. Species and speciation in phylogenetic systematics, with examples from the North American fish fauna. Ann. Mo. Bot. Gard. 72:596–635.

Williams, W. T., and H. T. Clifford. 1971. On the comparison of two classifications of the same set of elements. Taxon 20:519–522.

Zandee, M., and R. Geesink. 1987. Phylogenetics and legumes: A desire for the impossible? pp. 131–167. In: Advances in Legume Systematics, vol. 3 (C. H. Stirton, ed.). Royal Botanical Gardens, Kew, England.

Zandee, M., and M. C. Roos. 1987. Component compatibility in historical biogeography. Cladistics 3:305–332.

ANSWERS TO EXERCISES

Two conventions are used in the exercises. For complicated trees, the topology is presented in parenthetical notation, and the synapomorphies are listed by group and the autapomorphies by terminal taxon. In a few cases, a tree is so complicated that we present the tree with labeled internodes. Characters supporting groupings are then shown by internode. IG refers to the ingroup. Characters for the IG refer to synapomorphies for an entire ingroup.

CHAPTER 2

EXERCISE 2.1.—Analysis of *Sus*.

Each specific epithet has been abbreviated to its first letter.

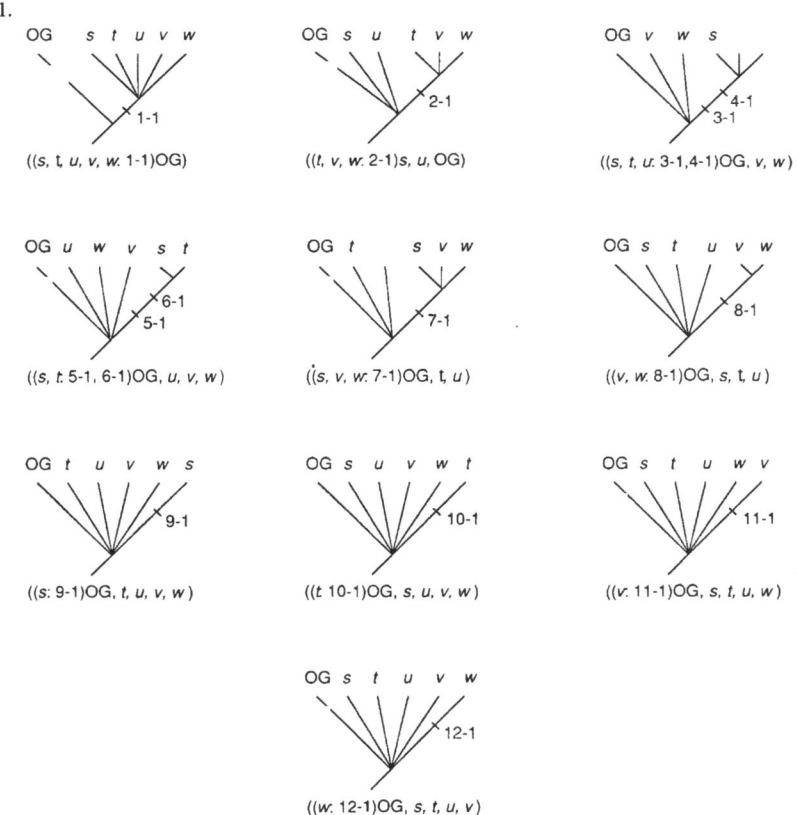

1.

((s, t, u, v, w: 1-1)OG)

((t, v, w: 2-1)s, u, OG)

((s, t, u: 3-1,4-1)OG, v, w)

((s, t: 5-1, 6-1)OG, u, v, w)

((s, v, w: 7-1)OG, t, u)

((v, w: 8-1)OG, s, t, u)

((s: 9-1)OG, t, u, v, w)

((t: 10-1)OG, s, u, v, w)

((v: 11-1)OG, s, t, u, w)

((w: 12-1)OG, s, t, u, v)

2. Topology: (OG(*vus, wus*)(*uus*(*sus, tus*))). Characters: (IG): 1-1; (*vus, wus*): 2-1; 7-1, 8-1; (*uus, sus, tus*): 3-1, 4-1; (*sus, tus*): 5-1, 6-1; (*sus*): 7-1, 9-1; (*tus*): 2-1, 10-1; (*vus*): 11-1; (*wus*): 12-1. ◉ = character showing convergence/parallelism or reversal (homoplasies).

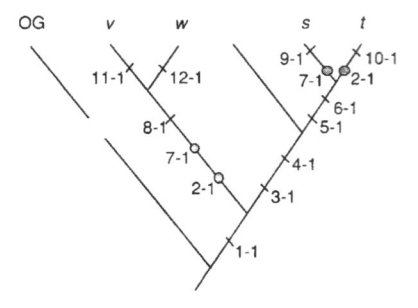

```
OG      v       w            s       t
 \       \       \          9-1\   /10-1
  \    11-1\  /12-1          7-1◉ ◉2-1
   \        \/                   /6-1
    \       /                   /5-1
     \   8-1/                  /
      \   7-1◉               /4-1
       \   2-1◉          /3-1
        \             /
         \          /
          \ 1-1  /
           \   /
            \ /
```

3. Tree statistics: CI (tree) = 0.857; length = 14. CI (characters): 1, 3–6, 8–12 = 1.0; 2, 7 = 0.5.

EXERCISE 2.2.—Analysis of Midae.

Each generic name has been abbreviated to its first letter.

1. Hypotheses of synapomorphy: (*M, N, O, Q, R*): 1-1; (IG): 2-1; (*P, Q, R*): 3-1; (*M P, Q, R*): 4-1; (*M, N, O*): 5-1, 10-1, 11-1; (*M, N, O, Q*): 6-1; (*N, O, Q, R*): 7-1; (*P*): 8-1; (*M*): 9-1.
2. Most parsimonious tree topology: (OG(*P*(*R*(*Q*(*M*(*N, O*)))))). Characters: (IG): 2-1, 3-1, 4-1; (*R Q, M, N, O*): 1-1, 7-1; (*Q, M, N, O*): 6-1; (*M, N, O*): 3-0, 5-1, 10-1, 11-1; (*N, O*): 4-0; (*P*): 8-1; (*M*): 7-0, 9-1.
3. Tree statistics: CI (tree) = $^{11}/_{14}$ = 0.786; length = 14. CI (characters): 1, 2, 5, 6, 8–11 = 1.0; 3, 4, 7 = 0.5.

EXERCISE 2.3.—Analysis of *Aus*.

Each specific epithet has been abbreviated to its first letter.

1, 2. There are two tree topologies and a total of four trees (two trees for each topology). All four trees share the following autapomorphies: (*a*): 6-1; (*c*): 10-1; (*d*): 8-1; (*e*): 9-1. Note that character 7-1 is homoplasious for both taxa *b* and *e* on all four trees.
Tree topology 1: (OG((*a, b*)*c*)(*d, e*)). Synapomorphies for tree 1A—(*a, b, c, d, e*): 1-1, 2-1; (*a, b, c*): 3-1; (*a, b*): 2-0, 5-1. Synapomorphies for tree 1B—(*a, b, c, d, e*): 1-1; (*a, b, c*): 3-1; (*a, b*): 5-1; (*d, e*): 2-1, 4-1. Autapomorphies unique for tree 1B—(*c*): 2-1. CI (characters): 1, 3–6, 8–10 = 1.0; 2, 7 = 0.5.
Tree topology 2: (OG(*a, b*)(*c*(*d, e*))). Synapomorphies for tree 2A—(*a, b, c, d, e*): 1-1, 3-1; (*a, b*): 5-1; (*c, d, e*): 2-1; (*d, e*): 3-0, 4-1. Synapomorphies for tree 2B—(*a, b, c, d, e*): 1-1; (*a, b*): 3-1, 5-1; (*c, d, e*): 2-1; (*d, e*): 4-1. Autapomorphies unique for tree 2B—(*c*): 3-1. CI (characters): 1, 2, 4–6, 8–10 = 1.0; 3, 7 = 0.5.
3. Tree statistics: CI (tree) = 0.833; length = 12.

CHAPTER 3

EXERCISE 3.1

Did you label node 1? If so, you're wrong. Node 1 is the root and you cannot label it without using its outgroup.

	Nodes					
TS	2	3	4	5	6	OG
3	a,b	b	b	a,b	b	decisive
4	a	a	b	a,b	a	decisive

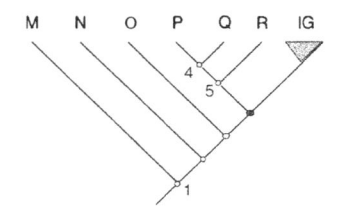

EXERCISE 3.2

		Nodes			
TS	T	U	V	W	OG
1		a,b	b	b	decisive
2	a	a	a,b	b	decisive
3	a	a	a	a,b	equivocal
4	a,b	a	a,b	a	decisive
5	a	a,b	b	a,b	equivocal
6	a,b	a	a,b	b	decisive

EXERCISE 3.3

		Nodes			
TS	M	O	P	Q	OG
1	a	a,b	b	b	decisive
2	a,b	b	b	b	decisive
3	a,b	a	a,b	a	decisive
4	a	a,b	a	a	decisive
5	a	a	b	a,b	equivocal
6		a,b	a,b	a,b	equivocal

EXERCISE 3.4

			Nodes				
TS	M	N	O	P	Q	R	OG
1	a	a,b	b	a,b	a		decisive
2	b	b	b	b	a,b		decisive
3	a,b	a	a,b	a	a		decisive
4	b	b	a,b	b	a,b		decisive
5	b	b	a,b	a	a		decisive
6	b	b	b	a,b	a		decisive

EXERCISE 3.5

TS	P	Q	R	S	T	U	V	W	X	Y	OG
1	a	b	a,b	a	a	b	b	a,b	a	a	decisive
2	a,b	a	a	a	a	b	b	b	b	a,b	equivocal
3	a	a	a	a,b	a	b	a,b	a	a	a	decisive
4	a,b	a,b	b	b	b	a,b	b	a,b	b	b	decisive
5	a	b	b	b	a,b	b	b	a,b	a	a	decisive
6	a	b	b	b	a,b	b	b	b	a,b	a,b	equivocal

Nodes

EXERCISE 3.6

1. Linear coding and additive binary coding.
2. Nonadditive and mixed are not applicable because the character tree is linear.
3. Note that the plesiomorphic character "1" is not listed because its vector is all zeros.

Taxon	Linear	Additive binary coding	
		m	n
OG	0	0	0	0
A	1	1	0	0
B	2		1	0
C	3			1

4. Tree: (OG(A(B, C))).

EXERCISE 3.7

1. Nonadditive binary coding.
2. Linear coding and additive binary coding cannot be used because the tree represents a branching TS. Mixed coding would not save many columns because the tree is symmetrical.
3.

Taxon	Nonadditive binary coding						
	m		o	p	q		
OG	0	0	0	0	0	0	0
A	1	0	0	0	0	0	0
B	0	0	0	0	1	0	0
C	0	0	0	0	0	0	0
D	0	0	0	0		1	0
E	0	0	0	1	1	0
F	0	0	0	1	0	1
S	1	0	0	0	0	0
T		0	1	0	0	0
U		0	1	0	0	0
V		1	0	0	0	0
W		0	0	0	0	0

4. Tree: (OG(A(B, C, F(D, E))(S(T, U)(V, W)))).

EXERCISE 3.8

1. Nonadditive binary and mixed coding; mixed coding is preferred because it would save columns.
2. Linear and additive coding will not work because the tree represents a branching TS.
3.

Taxon	Nonadditive binary coding								Mixed coding			
	n	o	p	q	r	s	t	u	n, q, s, t	r	u	o, p
OG	0	0	0	0	0	0	0	0	0	0	0	0
A	1	0	0	0	0	0	0	0	1	0	0	0
B	1	1	0	0	0	0	0	0	1	0	0	1
C	1	1	1	0	0	0	0	0	1	0	0	2
D	1	1	1	0	0	0	0	0	1	0	0	2
M	1	0	0	1	0	0	0	0	2	0	0	0
N	1	0	0	1	1	0	0	0	2	1	0	0
O	1	0	0	1	0	1	0	0	3	0	0	0
P	1	0	0	1	0	1	0	1	3	0	1	0
Q	1	0	0	1	0	1	1	0	4	0	0	0
R	1	0	0	1	0	1	1	0	4	0	0	0

4. Tree: (OG(A(B(C, D))(M, N(O, P(Q, R))))).

CHAPTER 4

EXERCISE 4.1.—Bremer's *Leysera* data.

1.

Taxon	Characters*					
	-	-	3	4	-	-
OG	0	0	0	0	0	0
L. longipes	0	0	0	0	0	0
L. leyseroides	1	1	1	1	1	1
L. tennella					1	1
L. gnaphalodes	1	1	1	1	0	0

* 1 = receptacle, 2 = floret tubules, 3 = pappus type, 4 = achene surface, 5 = pappus scales, 6 = life cycle.

2. Specific epithets are abbreviated. Tree: (OG(*lo*(*gn*(*ten*, *ly*)))). Characters within *Lysera*: (*ly*, *ten*, *gn*): 1-1, 2-1, 3-1, 4-1; (*ten*, *ly*): 5-1, 6-1. Synapomorphies of *Lysera*: $2N = 8$, solitary capitula on long peduncle.

EXERCISE 4.2.—Siegel-Causey's cliff shags.

1. OG vector: 1-0, 2-0/1, 22-0, 36-0/1, 39-0, 40-1, 42-0, 48-0, 63-0, 69-0, 78-0, 79-1, 81-0, 94-0, 97-0, 100-1, 102-0/1, 110-0, 111-0, 112-0, 114-0, 120-0, 124-0/1, 131-0, 134-0.

2. We did not consider TS 2, TS 36, TS 102, and TS 124 because their character decisions were equivocal at the OG node.

Some specific epithets are abbreviated. Tree: (OG((*pelag*, *urile*)(*arist*(*gaim*(*punct*, *feath*))))).

Characters: (IG): 97-1; (*pelag*, *urile*): 134-1; (*arist*, *gaim*, *punct*, *feath*): 63-1, 110-1; (*gaim*, *punct*, *feath*): 39-1, 40-0, 79-0, 94-1, 111-1, 112-1; (*punct*, *feath*): 1-1, 78-1; (*pelag*): 42-1; (*urile*): 69-1, 120-1, 131-1; (*arist*): 22-1, 100-0; (*gaim*): 48-1, 114-1; (*punct*): none; (*feath*): 81-1.

3. TS 2 and TS 102 remain equivocal. Character 36-0 is a synapomorphy shared by *gaim, punct,* and *feath.* Character 124-0 is either a homoplasy shared by *gaim* and *feath* or a synapomorphy below *gaim* that reverses in *punct.*

EXERCISE 4.3

$D(A,B) = \sum |X(A,i) - X(B,i)|$
$ |0-0| + |1-1| + |0\ \ 1| + |1-0| + |1\ \ 0|$
$ 3$
$D(M,B) = 3$

EXERCISE 4.4

1. $D(M,A) = \sum |X(M,i) - X(A,i)|$
   $ = |1-0| + |0-1| + |0\ \ 0| + |0-0| + |0\ \ 0|$
   $ 2$
   $D(M,B) = 3$

	ANC	B
ANC	–	
A	2	
B	3	3
C	4	4	–

2. $INT(A) = D[A, ANC(A)]$
   $ = 2$

EXERCISE 4.5.—Bremer's *Leysera* data.

1. Construct a data matrix.

Taxon	Transformation series					
	·	–	3	4		
Ancestor (ANC)	0	0	0	0	0	0
longipes (ln)	0	0	0	0	0	0
leyseroides (ly)	1	1	1	1	1	1
tennella (te)					1	1
gnaphoides (gn)	1				0	0

2. Calculate initial differences.

   $D(ANC, ln) = 0$
   $D(ANC, ly) = 6$
   $D(ANC, te) = 6$
   $D(ANC, gn) = 4$

ANSWERS TO EXERCISES

3. Add the taxon (*ln*) with least difference to ancestor (ANC).

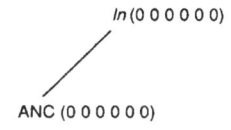

```
           ln(0 0 0 0 0 0)
          /
         /
        /
ANC (0 0 0 0 0 0)
```

4. Add *gnaphaloides*.

	Transformation series					
Taxon	1	2	3	4	5	6
ANC	0	0	0	0	0	0
gn	1	1	1	1	0	0
ln	0	0	0	0	0	0
A	0	0	0	0	0	0

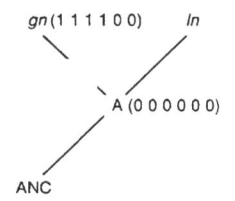

```
gn(1 1 1 1 0 0)      ln
      \             /
       \           /
         A(0 0 0 0 0 0)
        /
       /
    ANC
```

5. Add either *ly* or *te*—let's use *ly*.

$$D(ly, ln) = 6$$

$$D(ly, gn) = 2$$

$$D(ly, A) = 6$$

$$D(ly, ANC) = 6$$

$$D(ly, INT(ln)) = \frac{D(ly, ln) + D(ly, A) - D(ln, A)}{2}$$

$$= \frac{6 + 6 - 0}{2}$$

$$6$$

$$D(ly, INT(gn)) = \frac{D(ly, gn) + D(ly, A) - D(gn, A)}{2}$$

$$= \frac{2 + 6 - 4}{2}$$

$$2$$

$$D(ly, INT(A)) = \frac{D(ly, A) + D(ly, ANC) - D(A, ANC)}{2}$$

$$= \frac{6 + 6 - 0}{2}$$

$$6$$

Conclusion: add *ly* to INT (*gn*).

6. Calculate a new ancestor and an intermediate tree.

Taxon	Transformation series					
	1	2	3	4	5	6
A	0	0	0	0	0	0
gn	1	1	1	1	0	0
ly	1	1	1	1	1	1
B	1	1	1	1	0	0

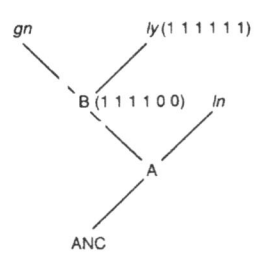

```
gn            ly(1 1 1 1 1 1)
  \          /
   \        /
    B (1 1 1 1 0 0)    ln
     \              /
      \            /
        A
        |
       ANC
```

7. Add *tennella* in the same manner.

$D(te,ly) = 0$ $D(ln,A) = 0$ $D[te\ \text{INT}(gn)] = 2$
$D(te,gn) = 2$ $D(B,A) = 4$ $D[te,\text{INT}(ly)] = 0$
$D(te,B) = 2$ $D(A,\text{ANC}) = 0$ $D[te,\text{INT}(B)] = 2$
$D(te,ln) = 6$ $D(ly,B) = 2$ $D[te,\text{INT}(ln)] = 6$
$D(te,A) = 6$ $D(gn,B) = 0$ $D[te,\text{INT}(A)] = 6$
$D(te,\text{ANC}) = 6$

8. Calculate a new ancestor and find the final tree.

Taxon	Transformation series					
	1	2	3	4	5	6
B	1	1	1	1	0	0
te	1	1	1	1	1	1
ly	1	1	1	1	1	1
C	1	1	1	1	1	1

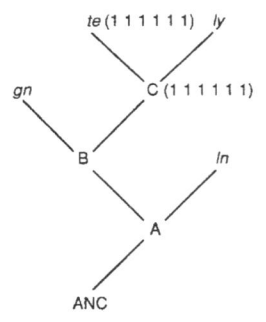

```
                te(1 1 1 1 1 1)   ly
                    \            /
                     \          /
gn                    C (1 1 1 1 1 1)
  \                  /
   \                /
    B              /
     \            / ln
      \          /
        A
        |
       ANC
```

EXERCISE 4.6.—Siegel-Causey's cliff shags.

1. Calculate differences.

$D(OG,pelag) = 4$ $D(OG,gaim) = 14$
$D(OG,urile) = 6$ $D(OG,punct) = 12$
$D(OG,arist) = 6$ $D(OG,feath) = 14$

Begin with *pelag*.

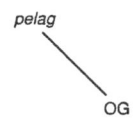

```
pelag
    \
     \
      OG
```

ANSWERS TO EXERCISES

2. The next shortest are *urile* or *arist*. We picked *arist* (but it won't make any difference at the end).

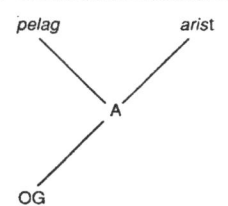

```
pelag           arist
    \          /
     \        /
      \  A  /
       \   /
        \ /
        /
       /
      /
    OG
```

	Transformation series
Taxon	1 2 22 36 39 40 42 48 63 69 78 79 81 94 97 100 102 110 111 112 114 120 124 131 134
OG	0 0 0 1 0 1 0 0 0 0 0 1 0 0 0 1 0 0 0 0 0 0 1 0 0
pelag	0 1 0 1 0 1 1 0 0 0 0 1 0 0 1 1 0 0 0 0 0 0 1 0 1
arist	0 0 1 1 0 1 0 0 1 0 0 1 0 0 1 0 1 1 0 0 0 0 1 0 0
A	0 0 0 1 0 1 0 0 0 0 0 1 0 0 1 1 0 0 0 0 0 0 1 0 0

3. Now pick *urile*.

$D[urile, INT(A)] = 5$ $D(urile, pelag) = 4$
$D(urile, A) = 5$ $D[urile, INT(arist)] = 5$
$D(urile, OG) = 6$ $D(urile, arist) = 10$
$D[urile, INT(pelag)] = 3$

Join *urile* to *pelag* by ancestor B. Calculate B.

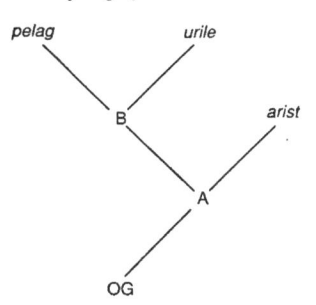

```
pelag         urile
    \        /
     \      /
      \ B /
       \ /          arist
        \          /
         \        /
          \  A  /
           \   /
           /
          /
         /
        OG
```

	Transformation series
Taxon	1 2 22 36 39 40 42 48 63 69 78 79 81 94 97 100 102 110 111 112 114 120 124 131 134
A	0 0 0 1 0 1 0 0 0 0 0 1 0 0 1 1 0 0 0 0 0 0 1 0 0
pelag	0 1 0 1 0 1 1 0 0 0 0 1 0 0 1 1 0 0 0 0 0 0 1 0 1
urile	0 1 0 1 0 1 0 0 0 1 0 1 0 0 1 1 0 0 0 0 0 1 1 1 1
B	0 1 0 1 0 1 0 0 0 0 0 1 0 0 1 1 0 0 0 0 0 0 1 0 1

4. Pick *punct* to add next.

D(*punct*,OG) = 12	D[*punct*,INT(*arist*)] = 9	D(*punct*,*arist*) = 12
D[*punct*,INT(A)] = 11	D(*punct*,A) = 11	D[*punct*,INT(B)] = 11
D(*punct*,B) = 13	D[*punct*,INT(*urile*)] = 13*	D(*punct*,*urile*) = 16*
D[*punct*,INT(*pelag*)] = 13*	D(*punct*,*pelag*) = 14*	

(*As per instructions, it is not necessary to calculate these distances because the distance of *punct* to INT(*arist*) is less than the distance between *punct* and INT(B). Therefore, there is no need to calculate distances between *punct* and terminal taxa above B.)

Add *punct* to *arist* by ancestor C. Calculate C.

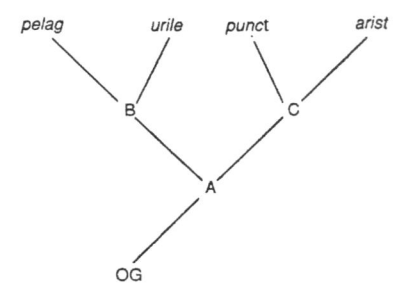

	Transformation series
Taxon 1	2 22 36 39 40 42 48 63 69 78 79 81 94 97 100 102 110 111 112 114 120 124 131 134
A	0 0 0 1 0 1 0 0 0 0 0 1 0 0 1 1 0 0 0 0 0 0 1 0 0
punct 1	0 0 0 1 0 0 0 1 0 1 0 0 1 1 1 0 1 1 1 0 0 1 0 0
arist 0	0 1 1 0 1 0 0 1 0 0 1 0 0 1 0 1 1 0 0 0 0 1 0 0
C	0 0 0 1 0 1 0 0 1 0 0 1 0 0 1 1 0 1 0 0 0 0 1 0 0

5. Pick *gaim* to add next.

D(*gaim*,OG) = 14	D[*gaim*,INT(A)] = 13	D(*gaim*,A) = 13
D[*gaim*,INT(B)] = 13	D(*gaim*,B) = 15	D[*gaim*,INT(C)] = 11
D(*gaim*,C) = 11	D[*gaim*,INT(*punct*)] = 4	D(*gaim*,*punct*) = 6
D[*gaim*,INT(*arist*)] = 10	D(*gaim*,*arist*) = 12	

Add *gaim* to *punct* through D. Calculate D.

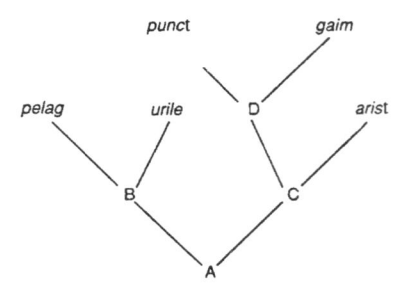

ANSWERS TO EXERCISES

	Transformation series
Taxon	1 2 22 36 39 40 42 48 63 69 78 79 81 94 97 100 102 110 111 112 114 120 124 131 134
C	0 0 0 1 0 1 0 0 1 0 0 1 0 0 1 1 0 1 0 0 0 0 1 0 0
punct	1 0 0 0 1 0 0 0 1 0 1 0 0 1 1 1 0 1 1 1 0 0 1 0 0
gaim	0 0 0 0 1 0 0 1 1 0 0 0 0 1 1 1 1 1 1 1 0 0 0 0 0
D	0 0 0 0 1 0 0 0 1 0 0 0 0 1 1 1 0 1 1 1 0 0 1 0 0

6. Finally (!) add *feath* to the tree.

$D(feath,OG) = 14$	$D[feath,INT(A)] = 13$	$D(feath,A) = 13$
$D[feath,INT(B)] = 13$	$D(feath,B) = 15$	$D[feath,INT(C)] = 11$
$D(feath,C) = 11$	$D[feath,INT(D)] = 4$	$D(feath,D) = 4$
$D[feath,INT(punct)] = 2$	$D(feath,punct) = 2$	$D[feath,INT(gaim)] = 3$
$D(feath,gaim) = 6$		

Add *feath* to *punct* through E. Calculate E.

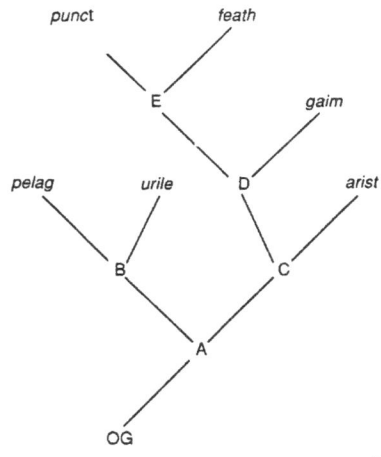

	Transformation series
Taxon	1 2 22 36 39 40 42 48 63 69 78 79 81 94 97 100 102 110 111 112 114 120 124 131 134
D	0 0 0 0 1 0 0 0 1 0 0 0 0 1 1 1 0 1 1 1 0 0 1 0 0
punct	1 0 0 0 1 0 0 0 1 0 1 0 0 1 1 1 0 1 1 1 0 0 1 0 0
feath	1 0 0 0 1 0 0 0 1 0 1 0 1 1 1 1 0 1 1 1 0 0 0 0 0
E	1 0 0 0 1 0 0 0 1 0 1 0 0 1 1 1 0 1 1 1 0 0 1 0 0

EXERCISE 4.7

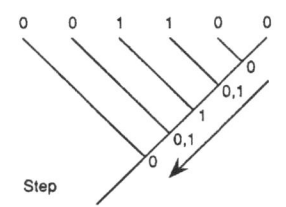

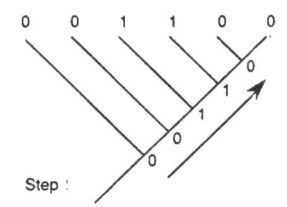

EXERCISE 4.8

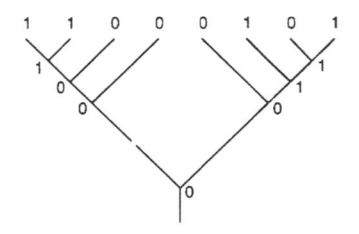

```
 1  1  0  0  0  1  0  1
  \ \ /  \ \ \  \ /
   1      0     1
    0      1
     \    /
      \  /
       0
       |
```

EXERCISE 4.9

```
1  0  0  1  0  0  1
 \ \ /  \ \ \ /
  0      0
         0
        0
       0
      0
```

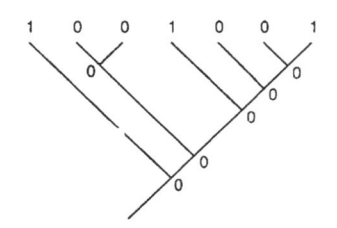

EXERCISE 4.10

DELTRAN Optimization

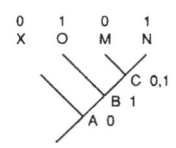

```
    0   1   0   1
    X   O   M   N
         \   \ / C 0,1
          \  B 1
           A 0
```

Step 1: Label ancestor nodes and assign root state. (This is a trivial task.)

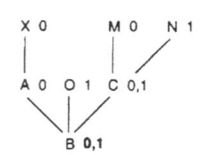

```
X 0        M 0   N 1
 \          \   /
  A 0  O 1  C 0,1
   \    \   /
    \    \ /
     B 0,1
```

Step 2: Reroot at B, the next ancestor up from the root, and recalculate the MPR. MPR(B) = (0,1).

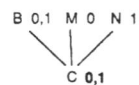

```
B 0,1   M 0   N 1
   \     \   /
    \    C 0,1
```

Step 3: Reroot at C, the last ancestor in the tree; recalculate the MPR. MPR(C) = (0,1).

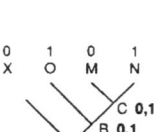

```
    0   1   0   1
    X   O   M   N
         \   \ / C 0,1
          \  B 0,1
           A 0
```

Step 4: Label all ancestor nodes with appropriate MPRs.

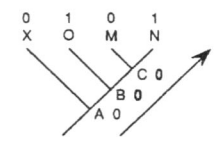

```
    0   1   0   1
    X   O   M   N
         \   \ / C 0
          \  B 0
           A 0
```

Step 5: Assign ancestor states on upward pass.

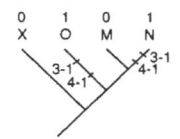

```
    0   1   0   1
    X   O   M   N
         3-1     3-1
         4-1     4-1
```

Step 6: Place character changes on appropriate branches following the state of the ancestral node.

ANSWERS TO EXERCISES

ACCTRAN Optimization

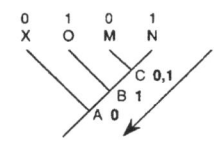

Step 1: Assign ancestral states on downward pass.

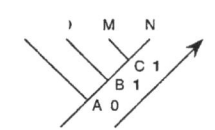

Step 2: Assign states on upward pass.

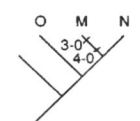

Step 3: Place character states on tree following states of ancestral nodes.

EXERCISE 4.11

DELTRAN Optimization

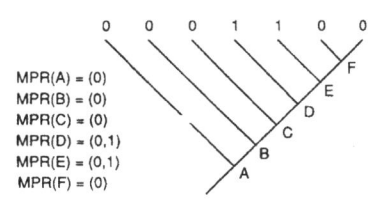

MPR(A) = (0)
MPR(B) = (0)
MPR(C) = (0)
MPR(D) = (0,1)
MPR(E) = (0,1)
MPR(F) = (0)

Step 1: Calculate MPRs for each ancestor node.

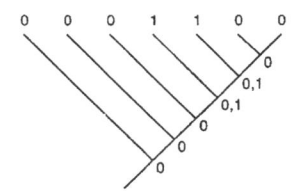

Step 2: Label each ancestor node with appropriate MPR.

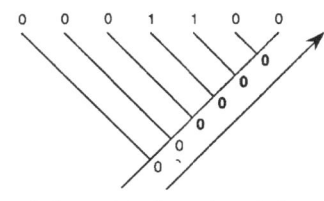

Step 3: Assign state of ancestor node by upward pass rules.

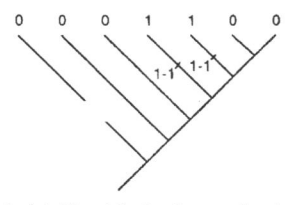

Step 4: Label tree following the ancestor character states.

ACCTRAN Optimization

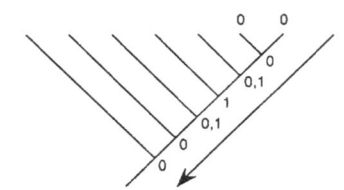

Step 1: Assign ancestor states by downward pass rules.

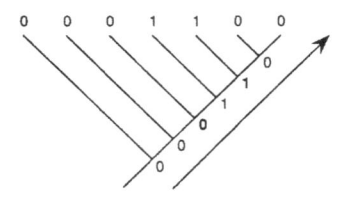

Step 2. Assign states of ancestor nodes by upward pass rules.

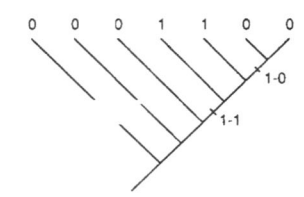

Step 3: Label tree following ancestor states.

CHAPTER 5

EXERCISE 5.1

Tree: (X(A(B, C))). M = 8, S = 8, G = 10, tree length = 8, CI = 1.0, R = 1.0, RC 1.0.

EXERCISE 5.2

M = 5, S = 6, G = 9, tree length = 6, CI = 0.833, R = 0.750, RC = 0.625.

EXERCISE 5.3

M = 8, S = 9, G = 11, tree length = 9, CI = 0.889, R = 0.667, RC = 0.593.

EXERCISE 5.4

Tree: (OG(O(M, N))). M = 10, S = 12, G = 15, tree length = 12, CI = 0.833, R = 0.600 RC = 0.500.

EXERCISE 5.5

Tree: (OG(A((B, C)((D, E)F)))). M = 11, S = 15, G = 26, tree length = 15, CI = 0.733, R = 0.733, RC = 0.538.

EXERCISE 5.6

Strict consensus tree: (A, B(C, D, E)).

EXERCISE 5.7

Adams consensus tree: (A(B, C, D)).

EXERCISE 5.8

Adams consensus tree: (A, B(C, D, E)).

EXERCISE 5.9

Majority consensus tree: ((W, V)(X(Y, Z))).

EXERCISE 5.10

Strict consensus tree: ((A, B, C)(D, E, F, G, H)).
Adams consensus tree: ((A, B, C)(D, E, H(F, G))).
Majority consensus tree: ((A, B, C)(D(E, F(G, H)))).

CHAPTER 6

EXERCISE 6.1.—Phylogeny of the Recent tetrapod vertebrates.

Classification a: (Lissamphibia, Aves, Mammalia(Chelonia, Lepidosauria, Crocodylia)). 45 trees possible. Classification is logically inconsistent with the phylogeny. "Reptilia" includes Chelonia, Lepidosauria, and Crocodylia.

Classification b: (Lissamphibia, Mammalia, Chelonia, Lepidosauria, Crocodylia, Aves). 945 trees possible. Classification is logically consistent with the phylogeny.

Classification c: Tree topology is identical to that of the tree in Fig. 6.10. One tree possible. Classification is logically consistent with the phylogeny.

EXERCISE 6.2.—Phylogeny of the land plants.

Tree: ((Anthoceropsida, Marchantiopsida, Bryopsida)(Psilotophytina, Lycopodophytina, Sphenophytina, Pteridophytina(Cycadopsida, Pinopsida, Ginkgopsida, Gnetopsida, Angiospermopsida))). Bryophyta includes Anthoceropsida, Marchantiopsida, and Bryopsida. Tracheophyta includes Psilotophytina, Lycopodophytina, Sphenophytina, Pteridophytina, and Spermatophytina. Spermatophytina includes Cycadopsida, Pinopsida, Ginkgopsida, Gnetopsida, and Angiospermopsida. 33,075 trees possible. Classification is logically inconsistent with the phylogeny.

EXERCISE 6.3

We have used each letter on the tree to stand for a genus. It doesn't matter if you picked different categories or different endings to the names. Instead, look for conformity in the *number* of categories. It shouldn't matter for systematic relationships if you have different names or endings.

1. Sequence convention:

    Order A–S
      Family A–F
        Genus *F*
        Genus *E*
        Genus *D*
        Genus *C*
        Genus *B*
        Genus *A*

Family M–S
  Genus *M*
  Genus *N*
  Genus *O*
  Genus *P*
  Genus *Q*
  Genus *R*
  Genus *S*

Hierarchical levels: 3.

2. All branch points named:

Order A–S
  Suborder A–F
    Family F
      Genus *F*
    Family A–E
      Subfamily E
        Genus *E*
      Subfamily A–D
        Tribe D
          Genus *D*
        Tribe A–C
          Subtribe C
            Genus *C*
          Subtribe AB
            Genus *A*
            Genus *B*
  Suborder M–S
    Family M
      Genus *M*
    Family N–S
      Subfamily N
        Genus *N*
      Subfamily O–S
        Tribe O
          Genus *O*
        Tribe P–S
          Subtribe P
            Genus *P*
          Subtribe Q–S
            Supergenus Q
              Genus *Q*
            Supergenus RS
              Genus *R*
              Genus *S*

Hierarchical levels: 8.

    3. Five hierarchical levels are saved by using the sequencing convention and not naming all the branch points.

ANSWERS TO EXERCISES

EXERCISE 6.4

We have used each letter on the tree to stand for a genus. It doesn't matter if you picked different categories or different endings to the names. Instead, look for conformity in the *number* of categories.

Order A–U
  Genus *N* (*incertae sedis*)
  Suborder M (*sedis mutabilis*)
    Genus *M*
  Suborder A–H (*sedis mutabilis*)
    Genus *H* (*incertae sedis*)
    Family G
      Genus *G*
    Family F
      Genus *F*
    Family A–E
      Subfamily C (*sedis mutabilis*)
        Genus *C*
      Subfamily D (*sedis mutabilis*)
        Genus *D*
      Subfamily E (*sedis mutabilis*)
        Genus *E*
      Subfamily AB
        Genus *A*
        Genus *B*
  Suborder P–U (*sedis mutabilis*)
    Family P
      Genus *P*
    Family Q
      Genus *Q*
    Family R
      Genus *R*
    Family S–U
      Genus *S* (*sedis mutabilis*)
      Genus *T* (*sedis mutabilis*)
      Genus *U* (*sedis mutabilis*)

EXERCISE 6.5

1. Order Tusiformes
    †Genus *X* (*incertae sedis*)
    Suborder A–F
      †Family F
        Genus *F*
      Family A–E
        †Subfamily E
          Genus *E*
        Subfamily A–D
          Tribe D
            Genus *D*

        Tribe A–C
            Genus *A* (*sedis mutabilis*)
            Genus *B* (*sedis mutabilis*)
            Genus *C* (*sedis mutabilis*)
        Suborder M–S
          Genus *P* (*incertae sedis*)
          Family M–O
            Tribe O
              Genus *O*
            Tribe MN
              Genus *M*
              †Genus *N*
          Family Q–S
            Tribe Q
              Genus *Q*
            Tribe RS
              Genus *R*
              Genus *S*
 2. Order Tusiformes
      †Genus *X* (*incertae sedis*)
      Suborder A–F
        Plesion F
        Plesion E
        Family D
          Genus *D*
        Family A–C
            Genus *A* (*sedis mutabilis*)
            Genus *B* (*sedis mutabilis*)
            Genus *C* (*sedis mutabilis*)
      Suborder M–S
        Genus *P* (*incertae sedis*)
        Family M–O
          Tribe O
            Genus *O*
          Plesion N
          Tribe M
            Genus *M*
        Family Q–S
          Tribe Q
            Genus *Q*
          Tribe R
            Genus *R*
          Tribe S
            Genus *S*

3. In #1, 6 categories and 29 entries are required in the classification; in #2, only 5 categories (a plesion doesn't count) and 26 entries. On a strictly numerical basis, use of Convention 7 is more justified than not using it. Use of plesions simplifies the classification without implying possibly unknown relationships.

EXERCISE 6.6

1.

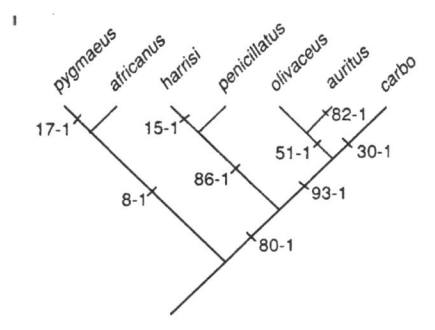

```
      pygmaeus  africanus  harrisi  penicillatus  olivaceus  auritus  carbo
         \        /          \          \           \         82-1    /
         17-1   15-1          \          \           \          \    /
            \   /              \         51-1         \        30-1
             \ /                \           \          \        /
              \                  86-1        \         93-1
              8-1                  \          \         /
                \                   \          \       /
                 \                   \          \_____/
                  _____/
                                      80-1
```

2.

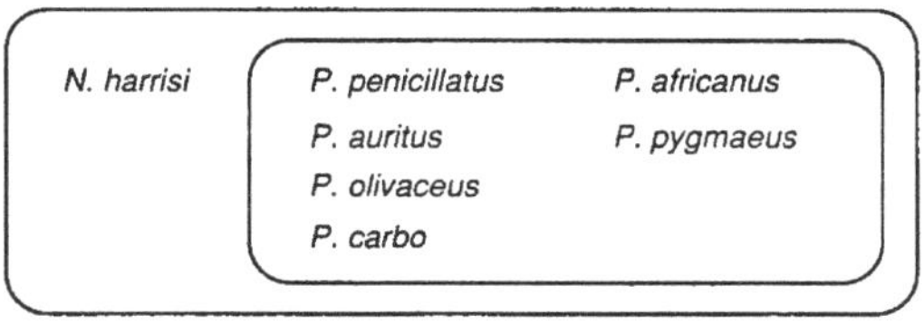

Venn diagram of traditional classification

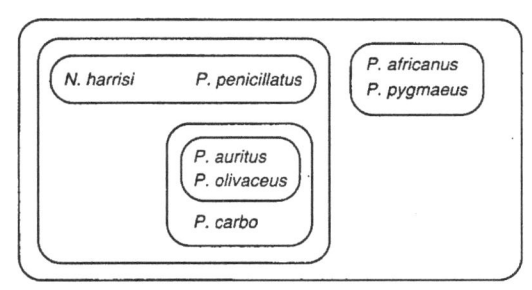

- N. harrisi
- P. penicillatus, P. auritus, P. olivaceus, P. carbo
- P. africanus, P. pygmaeus

Venn diagram of phylogeny

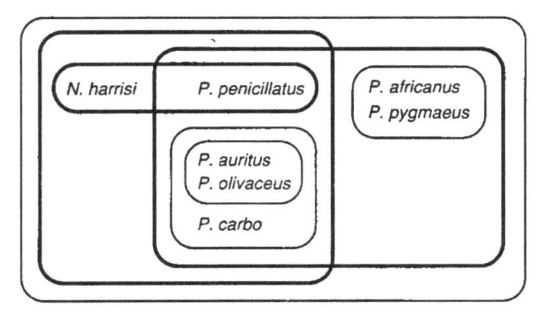

- Outer: N. harrisi, P. penicillatus, { P. auritus, P. olivaceus, P. carbo }
- P. africanus, P. pygmaeus

Venn diagram of consensus

- N. harrisi | P. penicillatus { P. auritus, P. olivaceus, P. carbo }
- P. africanus, P. pygmaeus

3. There are many classifications depending upon the nature of the categories you wish to use. Here is one, using the listing convention.

Firstgenus
 F. pygmaeus
 F. africanus
Secondgenus
 S. penicillatus
 S. harrisi
Thirdgenus
 T. carbo
 T. auritus
 T. olivaceus

CHAPTER 7

EXERCISE 7.1.—*Luxilus zonatus* species group.

1. Taxon tree: ((*cardinalis, pilsbryi*)*zonatus*). Ancestral taxa are labeled x and y.

Area	zonatus	cardinalis	pilsbryi		
1	0	1	0	1	
2	0	0	1	1	
3	1	0	0	0	

2. Area tree: ((1, 2)3).

EXERCISE 7.2.—*Fundulus nottii* species group.

Each of the specific epithets used in this exercise have been abbreviated to their first three letters. W, X, Y, and Z are ancestral taxa.

Area	LIN	ESC	NOT	W	X	BLA	DIS	Y	Z
1	1	0	0	0	1	0	0	0	1
2	0	1	0	1	1	0	0	0	1
3	0	0	1	1	1	0	0	0	1
4	0	0	0	0	0	1	0	1	1
5	0	0	0	0	0	0	1	1	1

Area tree: ((1(2,3))(4,5)). Trees are congruent.

EXERCISE 7.3.—Some moths.

1

	Taxa				
Area	M1	M2	Mx	M3	My
AS	1	0	1	0	
NG	0	1	1	0	1
SA	0	0	0	1	1
AF	0	0	?	0	?

2. One tree: (((AS, NG)SA)AF).
ACCTRAN optimization. Synapomorphies—(AS, NG): Mx-1; (AS, NG, SA): Bx-1, Wy-1; (AS, NG, SA, AF): Bz-1, Wz-1, My-1. Autapomorphies—(AS): B1-1, W1-1, M1-1; (NG): B2-1, M2-1; (SA): M3-1, W3-1; (AF): B4-1, W4-1.
DELTRAN optimization. Synapomorphies—(AS, NG): Mx-1, Bx-1; (AS, NG, SA): My-1, Wy-1; (AS, NG, SA, AF): Bz-1, Wz-1. Autapomorphies identical to those of ACCTRAN.

EXERCISE 7.4.—Ferns.

1

	Taxa				
Area	Fe12	Fe3	Fex	Fe4	Fey
AS	1	0		0	.
NG	1	0	1	0	
SA	0	1	1	0	
AF	0	0	0	1	

2. DELTRAN optimization. Tree 1: (((AS, NG)SA)AF). Synapomorphies—(AS, NG): Fe12-1, Tx-1, Fix-1; (AS, NG, SA): Fex-1; (AS, NG, SA, AF): Ty-1, Fiy-1, Fey-1. Autapomorphies—(AS): T1-1, Fi14-1; (NG): T2-1, Fi2-1; (SA): T34-1, Fi3-1, Fe3-1.
Tree 2: (((AS, NG)AF)SA). Synapomorphies—(AS, NG): Fe12-1, Tx-1, Fex-1; (AS, NG, AF): Fix-1; (AS, NG, AF, SA): Fey-1, Fiy-1, Ty-1. Autapomorphies—(AS): Fi14-1, T1-1; (NG): Fi2-1, T2-1; (AF): T34-1, Fe4-1, Fi14-1; (SA): T34-1, Fi3-1, Fe3-1, Fex-1.
Tree 3: ((AS, NG)(SA, AF)). Synapomorphies—(AS, NG): Tx-1, Fix-1, Fe12-1, Fex-1; (SA, AF): T34-1; (AS, NG, AF, SA): Ty-1, Fiy-1, Fey-1. Autapomorphies—(AS): T1-1, Fi14-1; (NG): T2-1, Fi2-1; (SA): Fi3-1, Fe3-1, Fex-1; (AF): Fi14-1, Fix-1, Fe4-1.
Tree statistics: CI = 0.833, length = 18, R = 0.500, RC = 0.417.

EXERCISE 7.5.—Combining the matrix.

The black dots are homoplasies. The arrow suggests moving ancestral taxon Fiy-1 up one branch. If ancestor Fiy-1 moves up, then ancestor Fix-1 should be removed from Africa because by placing ancestor Fiy-1 higher it can no longer give rise to Fix-1 in Africa. The higher position of Fiy-1 also suggests that Fi14-1 dispersed to Africa. The T34-1 distribution may represent dispersal or persistence of a widespread species. If dispersal, the direction cannot be determined.

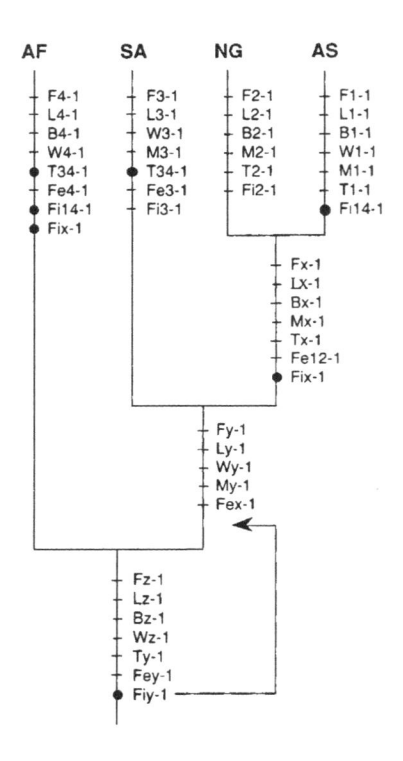

```
     AF          SA          NG          AS
    ┼ F4-1      ┼ F3-1      ┼ F2-1      ┼ F1-1
    ┼ L4-1      ┼ L3-1      ┼ L2-1      ┼ L1-1
    ┼ B4-1      ┼ W3-1      ┼ B2-1      ┼ B1-1
    ┼ W4-1      ┼ M3-1      ┼ M2-1      ┼ W1-1
    ● T34-1     ● T34-1     ┼ T2-1      ┼ M1-1
    ┼ Fe4-1     ┼ Fe3-1     ┼ Fi2-1     ┼ T1-1
    ● Fi14-1    ┼ Fi3-1                 ● Fi14-1
    ● Fix-1
                                ┼ Fx-1
                                ┼ Lx-1
                                ┼ Bx-1
                                ┼ Mx-1
                                ┼ Tx-1
                                ┼ Fe12-1
                                ● Fix-1

                    ┼ Fy-1
                    ┼ Ly-1
                    ┼ Wy-1
                    ┼ My-1
                    ┼ Fex-1 ←────┐
                                 │
                ┼ Fz-1            │
                ┼ Lz-1            │
                ┼ Bz-1            │
                ┼ Wz-1            │
                ┼ Ty-1            │
                ┼ Fey-1           │
                ● Fiy-1 ──────────┘
```

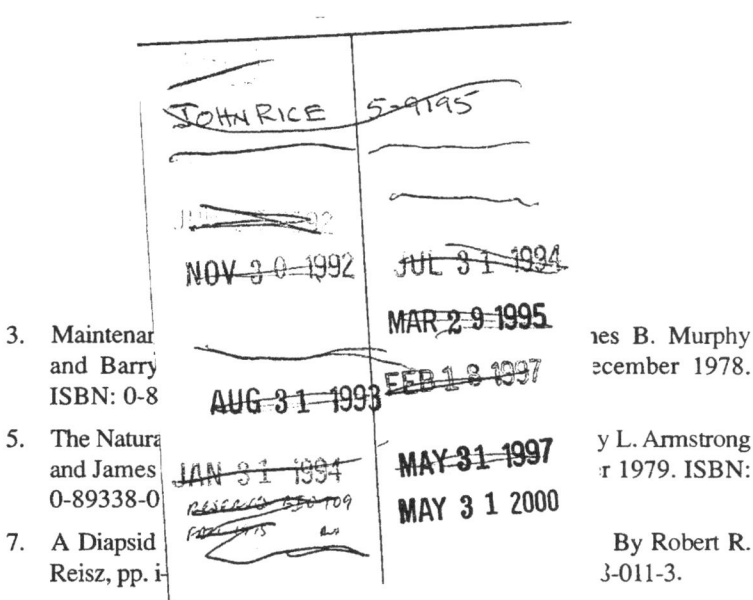

3. Maintenar es B. Murphy
and Barry ecember 1978.
ISBN: 0-8

5. The Natura y L. Armstrong
and James r 1979. ISBN:
0-89338-0

7. A Diapsid By Robert R.
Reisz, pp. i 3-011-3.

9. The Ecological Impact of Man on the South Florida Herpetofauna. By Larry David Wilson and Louis Porras, pp. i–vi, 1–89, 8 August 1983. ISBN: 0-89338-018-0.

10. Vertebrate Ecology and Systematics: A Tribute to Henry S. Fitch. Edited by Richard A. Seigel, Lawrence E. Hunt, James L. Knight, Luis Malaret, and Nancy Zuschlag, pp. i–viii, 1–278, 21 June 1984. ISBN: 0-89338-019-0.

13. Geographic Variation among Brown and Grizzly Bears (*Ursus arctos*) in North America. By E. Raymond Hall, pp. i–ii, 1–16, 10 August 1984.

15. Spring Geese and Other Poems. By Denise Low, pp. 1–84, September 1984. ISBN: 0-89338-024-5.

18. A Checklist of the Vertebrate Animals of Kansas. By George D. Potts and Joseph T. Collins, pp. i–vi, 1–42, September 1991. ISBN: 0-89338-038-5.

Lightning Source UK Ltd.
Milton Keynes UK
UKOW06f0216291015

261573UK00001B/107/P